363.11962233819

BP and the Macondo Spill

BP and the Macondo Spill

The Complete Story

Colin Read
Professor of Economics and Finance, SUNY
College, Plattsburgh, USA

palgrave
macmillan

First published 2011 by
PALGRAVE MACMILLAN

Palgrave Macmillan in the UK is an imprint of Macmillan Publishers Limited, registered in England, company number 785998, of Houndmills, Basingstoke, Hampshire RG21 6XS.

Palgrave Macmillan in the US is a division of St Martin's Press LLC, 175 Fifth Avenue, New York, NY 10010.

Palgrave Macmillan is the global academic imprint of the above companies and has companies and representatives throughout the world.

Palgrave® and Macmillan® are registered trademarks in the United States, the United Kingdom, Europe and other countries.

ISBN: 978–0–230–29358–8 hardback

This book is printed on paper suitable for recycling and made from fully managed and sustained forest sources. Logging, pulping and manufacturing processes are expected to conform to the environmental regulations of the country of origin.

A catalogue record for this book is available from the British Library.

A catalog record for this book is available from the Library of Congress.

10 9 8 7 6 5 4 3 2 1
20 19 18 17 16 15 14 13 12 11

Printed and bound in Great Britain by
CPI Antony Rowe, Chippenham and Eastbourne

Contents

Illustrations

Figures

Tables

About the Author

Colin Read is Professor of Economics and Finance, former dean of the School of Business and Economics at SUNY College, Plattsburgh, and a columnist for the *Plattsburgh (New York) Press-Republican* newspaper. He has a PhD in economics, a JD in law, an MBA, a masters in taxation, and has taught environmental and energy economics and finance for 25 years. Colin's recent books include *Global Financial Meltdown: How We Can Avoid the Next Economic Crisis, The Fear Factor, The Rise and Fall of an Economic Empire*, and a book on international taxation. He has written dozens of papers on market failure and volatility, and housing markets. He writes a weekly column in the *Plattsburgh Press-Republican* newspaper, and appears monthly on a local PBS television show to discuss the regional and national economy. He has worked as a research associate at the Harvard Joint Center for Housing Studies and served the Ministry of Finance in Indonesia under contract from the Harvard Institute for International Development. His consulting company can be found on the Internet at www.economicinsights.net. In his spare time he enjoys floatplane flying from his home on Lake Champlain that he shares with his wife, Natalie, daughter, Blair, and dog, Albert.

1
Timeline for the Deepwater Horizon Oil Spill

2008

March 19 – Macondo Prospect lease purchased by a consortium of BP (65%), Anadarko Petroleum (25%), and Mitsui Oil Exploration (10%). This lease sale included 614 other prospect leases, of which 34.5% were at the depth of the Macondo Prospect or deeper.[1]

2009

October 7 – BP began drilling at the MC252 Macondo Prospect site using the Transocean Marianas rig.
November 29 – Marianas rig damaged by Hurricane Ida; Deepwater Horizon brought in as a replacement.

2010

February – Transocean Deepwater Horizon resumes drilling.

March 8 – A stuck drill pipe requires BP to amend drill plan.

April 9 – Drilling completed and hole measurements commenced.

April 14–18 – Hole measurements used to optimize cement design.

April 19 – Cement job begins.

April 20 – 1:00 a.m. – Cement job completed.

April 20 – 9:50 p.m. – A well blowout on the rig caused a series of explosions and fire on the Deepwater Horizon rig. Eleven workers died.

April 22 – After two days ablaze, and millions of tons of water sprayed onto rig from fireboats, the rig topples and sinks to the bottom of the Gulf of Mexico. In the process, it severs the well riser pipe and initiates the Deepwater Horizon oil spill. The value of BP shares begins to plummet.

April 25 – The first estimate of a leak rate from the U.S. Coast Guard reports 1,000 barrels are escaping into the Gulf of Mexico each day. The estimate was soon upped to 5,000 barrels per day. BP begins to use underwater Remotely Operated Vehicles (ROVs) to shut down the blowout preventer.

April 29 – President Obama promises to hold BP responsible for the spill, and pledges all necessary resources to clean it up.

April 30 – BP Chief Executive Officer Tony Hayward accepts corporate responsibility for the spill and promises to pay for the cleanup and for all "legitimate costs."

May 2 – BP begins drilling the first of two relief wells to stop the spill. The U.S. government imposes a fishing ban for certain areas in the Gulf of Mexico.

May 7 – The first containment dome is lowered over the blowout preventer and bent riser in an effort to slow down the spill as BP proceeds to drill the relief well.

May 11 – Congress commands executives from the major oil companies that drill in the Gulf of Mexico (Big Oil) to testify in Washington, D.C.

May 14 – President Obama complains that Big Oil's testimony was an embarrassing exercise in finger pointing.

May 19 – The first oil slicks begin to hit Louisiana wetlands.

May 26 – A top kill designed to pump rubber, steel balls, and mud into the top of the well is initiated. After three days, BP declares the attempt a failure. The value of BP falls another 17% on announcement of the failure.

May 27 – Government-commissioned scientists up their spill-rate estimate to 12,000–19,000 barrels per day.

May 28 – The administration imposes a six-month moratorium on deepwater exploratory drilling.

June 1 – The Justice Department initiates a criminal and civil investigation of the spill.

June 2 – Fishing restrictions expanded to cover almost 2/5ths of federal waters in the Gulf of Mexico.

June 3 – ROVs cut off bent riser top as a step toward designing a tighter seal.

June 10 – With riser more cleanly cut off, scientists estimate spill rate at 20,000–40,000 barrels per day.

June 15 – Congress opens hearings to testimony of oil executives in an investigation of the spill.

June 16 – BP and the Obama administration agree to the creation of a $20 billion economic damages trust fund. Dividends to shareholders are suspended for the balance of the year.

June 17 – BP CEO Hayward receives an icy reception as he offers a stilted testimony to Congress.

June 18 – Anadarko accuses its partner BP of reckless behavior.

June 22 – Hayward cedes day-to-day management of the spill to Bob Dudley. Federal judge Feldman lifts the drilling moratorium.

July 12 – BP designs and installs a better fitting cap to reduce the oil spill.

July 15 – BP is able to shut down all oil flow.

July 20 – BP reports a $17 billion quarterly loss as it writes off costs of the spill.

August 2 – The government's Flow Rate Technical Group estimates that the spill released 4.1 million barrels.

August 4 – Scientists report all but approximately 25% of the spill remains unrecovered, unburned, or unevaporated.

August 5 – BP pumps cement into the top of the well to complete a static top kill.

September 3 – The damaged blowout preventer is detached by ROVs and brought to the surface to be impounded by the Department of Justice.

September 8 – BP releases its internal investigation of engineering and drilling errors that contributed to the spill.

September 14 – Judge Carl Barbier begins hearings to consolidate lawsuits in New Orleans.

September 16 – BP's relief well intercepts the wellbore. Drillers from Boots and Coots initiate bottom kill.

September 19 – Retired Admiral Thad Allen declares the well dead.

September 30 – Retired Admiral Thad Allen turns over his oversight of the cleanup.

October 12 – The Obama administration lifts the replacement moratorium on offshore drilling earlier than expected, but with expanded conditions for spill avoidance and readiness.

October 28 – President Obama's National Commission on the BP Deepwater Horizon Oil spill and Offshore Drilling investigatory committee issues a letter asserting that Halliburton failed to test the cement design it had employed in the well, and that this failure likely contributed to the well blowout. Halliburton's stock price fell significantly, as BP's stock price rose.

January 12, 2011 – President Obama's National Commission on the BP Deepwater Horizon Oil spill and Offshore Drilling investigatory committee releases its final report.

2
Introduction

We drive to the gas station to purchase fuel for our automobile. Heating oil is delivered automatically to our home to keep us warm through the winter. We do not stop to think about the raw materials that go into the plastics that make so many of the goods we consume. Nor do we dwell on the amount of fuel used to cultivate our food or bring it to market. As a matter of fact, we think little about energy until we are shocked by the displacements and damage caused by a major energy industry calamity. Unfortunately, when such a calamity occurs, we then seek simple explanations, despite our complicity in our increasingly desperate demand for energy.

An event such as the Deepwater Horizon oil well blowout and spill in the Gulf of Mexico should give us a chance to pause and try to make sense of the impact of an environmental tragedy. However, if we merely brush off such a tragedy as the result of carelessness of foreign Big Oil, we miss an opportunity to truly understand an exceedingly complex energy environment. And, by blaming a single entity, we avoid broader culpability in an industry that may need some reform, a governmental regulatory body that failed to protect the public, and a set of technologies that have lulled us into energy complacency. We do ourselves no justice by turning a perfect storm of unfortunate events into an oversimplified opportunity to partition blame disproportionately.

In this book, I will attempt to understand precisely what occurred during this most complex saga. While the primary goal is to explain the various factors that contributed to an oil-damaged gulf and a financially ravaged BP, the underlying theme is to better understand our increasing energy dependency. To fully grasp the role of BP, the British corporation formerly known as British Petroleum, we must describe a much more expansive context.

I begin in Part I with the natural and economic history of oil and its future. I document the source of oil, the ways we have extracted and used it since its widespread adoption as our primary energy source, and the increasing risk we are undertaking as oil is becoming scarcer.

In Part II, I provide a brief history of oil spills. I describe the dirty dozen spills that predated the Deepwater Horizon spill; I then separately treat the Exxon Valdez spill, the tragedy that changed regard for Big Oil in the United States. We will see that the planet has experienced many large spills, on land, the ocean surface, and deep below the ocean surface. These spills occurred in more and less sensitive ecosystems, were larger and smaller than the Deepwater Horizon spill, and lasted briefly or over decades.

In the third part, I describe the development of the Macondo oil reservoir and the events that led up to the fire and spill at the Deepwater Horizon oil drilling platform. I discuss the technologies of offshore and deepwater drilling, and their concomitant risks. I also discuss the organizational failures and risk management realities that demonstrate there can be no easy solution to our complicated demand for oil.

In Part IV, I explain why this first mammoth oil spill in U.S. territory in the era of 24 hour news cycles piqued the interest of a worldwide audience. I document how BP managed the crisis in a way that is likely more responsible and complete than any past transgressors, and yet took significant missteps and was pilloried in the media. I explain that, in the interest of telling a compelling story, the media may have erred in telling people what they wanted to hear or what would most rile the public and, in turn, cast to the wind certain journalistic ethics.

One of the lessons from the analysis is that it is dangerous to oversimplify complex situations. In this part, I also describe the science of what went horribly wrong for BP in April of 2010. While the media, the political machine, and financial markets were churning, BP was quietly, and some would say ploddingly, devising, testing, and then implementing a solution to the ruptured well. I document the capping and abatement processes, from the technological and economic to the human and ecosystem perspectives.

I then make a pronounced shift in Part V. I move from the rationality, and perhaps naivety, of the engineer to the politics of Congress. In any disaster, the public wants answers and decisiveness, and politicians are often eager to make the pronouncements that placate their public. I describe the interplay between politics and the courts as I summarize the complicated legal quagmire from which BP will struggle for years to extricate itself. In this part, I also discuss the role of a special master to administer economic claims on BP, and the talk of criminal and civil proceedings.

While this book was completed at a time when we likely fully understood the human, ecosystem, and financial obligations, U.S. oil spill history tells us that it will take decades to navigate the legal path this saga will take. I will describe the legal issues that will be invoked over the coming years, and postulate the ultimate success of various legal theories.

As the saga played out, financial markets incorporated every morsel of information and emotion into a BP stock price that was likened to a falling dagger – one tries to catch it at his peril. While I document the effects that should rationally influence the price of a stock, I demonstrate how rationality gave way to market emotionality.

I end the book with a summary of lessons the world can learn. There is easy oil no more. We as an energy consuming society will have lost an opportunity if we do not use this chance to reflect on our collective role that created the environment for a mega-spill waiting to happen. Certainly, if it were not BP, it would have been, and someday will be, another company. That is certain. What is less certain is whether we are willing to pay the price of episodic environmental degradation in our insatiable quest for cheap energy.

My interest in the BP episode arose from an undergraduate degree in physics that had me intrigued over the engineering challenges of the BP solution. I subsequently completed a PhD in economics, a Juris Doctor in law and a Master's of Business Administration. I have since been a dean of a business school in the State University of New York (SUNY) system, and now teach energy economics, environmental economics, and finance. I draw upon an appreciation for both the business and the legal aspects of this most legally, socially, and economically diverse case.

My goal is to provide you with a primer on the energy, legal, business, financial, and technological aspects of this environmental tragedy, from both an academic and practical perspective. I also hope to challenge you to become a participant in our collective energy future. My previous books, *Global Financial Meltdown*, *The Fear Factor*, and *The Rise and Fall of an Economic Empire*, share with this book the assumption that an economically literate citizenry is the best antidote to episodic and sensational economic reporting. If recent history has proven anything, it is that economic and environmental issues are very complex. The intricacies and subtleties of our economic and energy policies can be understood only to the extent of the sophistication of those reporting such important issues to us. Unfortunately, few reporters are economists, engineers, or environmentalists. Unless we trust their understanding and reinterpretation of complex economic and environmental phenomena, it is incumbent upon us to educate ourselves so we can judge and place in perspective the best reporting from the rest.

As this book winds down, I believe you will agree that our almost unwavering focus on BP throughout this environmental tragedy was misplaced. While a nation was justifiably angry and frustrated with such a tragedy, we must accept responsibility too. I am a consumer of energy products, and I have, at times, held BP in my retirement mutual funds and in stocks. Indeed, 40% of BP is owned by U.S. residents. It is not so simple to focus our frustrations on a single company when there is so much responsibility to go around. And, in this case, what goes around also comes around.

There is one thing for which we can all be assured. While the Deepwater Horizon fire and spill was the first deep-sea drilling-related spill to capture international attention and 24 hour scrutiny, our thirst for oil should tell us it will not be our last.

Part I

The Natural and Economic History of Oil

Oil is a powerful three-letter word. It connotes black gold, a liquid that will fuel our economy, a commodity for which many of us can quote its spot price, and a goo that damages our environment. In this first part, I discuss the natural and economic past, present, and future of oil. This discussion frames the question, "Why do we choose to take the inevitable risks that are necessary to quench our energy thirst?"

3
A Brief Natural History of Oil

Our fascination with oil is well-founded. Oil is now intertwined into our economy, our livelihood, and, increasingly, in our precarious energy future. The natural history of oil punctuates the earth's natural history, dating back to an era not unlike our own – one of rising carbon dioxide in our atmosphere.

Almost a century ago, an article linked the death of the dinosaur to the fuel in one's car.[2] In doing so, the article created a myth that has held broad acceptance ever since. However, scientists and geologists now agree that oil most likely formed not from the decomposition of large extinct land-based animals as was once thought, but rather from the aggregation of large stocks of some of the smallest sea-based organisms. Indeed, one of the best predictors of oil is in the identification of geological formations that could best harbor these marine sediments.

All organisms are made up of carbon-based molecules. Hydrogen and oxygen are the most common atoms to combine with carbon. Longer molecules that constitute sophisticated organic materials combine many other elements, in smaller amounts. These include sulfur, sodium, potassium, phosphorous, iron, magnesium, and many others.

Organisms have evolved to create longer, even more complex, and more specialized molecules that are constituted by the same elements that make up alkanes, the primary molecules in petroleum. The simplest of these organisms are microscopic, and most of them have resided in the sea through the prehistory of our planet. The combination of abundant water, sunlight, and atmospheric carbon dioxide provided the basic building blocks for these first simple organisms.

The availability of carbon dioxide, in addition to the plentiful sunlight and water, created the conditions for life, and ultimately for oil. As these organisms lived and grew 300 million years ago in the Paleozoic Era, they combined with water and high concentrations of carbon dioxide to produce hydrocarbons and sugars. As organisms died, gravity took them to the bottom of the ocean, where billions and billions of dead organisms amassed to produce thick layers of decomposing organic matter.

When geological processes cover up this layer of organic material with layers of sand, and then rock, the process to create oil is in place.

Under this prevailing theory of oil production, we can predict the discovery of oil based on a few precursors. The initial conditions are abundant sunlight and water, combined with the carbon dioxide upon which organisms thrive, forces that allow organisms to sink and form layers, and the creation of a blanketing layer of inorganic matter. Over time, this blanketing layer of sediment and sand turns to sedimentary rock. This blanket creates the high temperature and pressure that will bake the organic matter into oil.

There are periods in the earth's past that better create the concentrations of carbon dioxide so necessary for prolific organic matter, from microscopic organisms in the ocean to leafy matter on land. We now know from ice samples that the concentration of atmospheric carbon dioxide has cycled significantly over the millennia. The researchers Branola, Raynaud, Lorius, and Barkov recently reported that the level of atmospheric carbon dioxide follows regular cycles that correspond to increased glaciation when the carbon dioxide level in the atmosphere is low, to global warming in eras with a high atmospheric carbon dioxide level.[3]

Over these historic cycles, carbon dioxide concentrations reach lows that hover around 190 parts per million by volume (ppmv), while high levels, conducive to the creation of the algae and organisms that make oil, approach 290 ppmv. The earth is currently on the increasing trend of one of these 100,000 year cycles, but with much higher levels of atmospheric carbon dioxide, nearing 400 ppmv.[4] This much higher concentration is widely believed to be accelerated by our intensive use of oil since the onset of the industrial revolution, and its attendant release of previously sequestered carbon to the atmosphere.

Periods of rising carbon dioxide stimulate the growth of microorganisms and plants fueled by carbon dioxide. This process creates what physicists and engineers call a negative feedback loop. The higher carbon dioxide levels stimulate those plants and organisms that grow by absorbing carbon dioxide. If these organisms subsequently die and decay, the carbon dioxide is released back to the environment. However, if the organisms become trapped in a sedimentary layer, the carbon is sequestered, only to be released when the resulting oil is burned and the carbon dioxide is returned to the atmosphere.

Our first clues to the factors that combine to make oil came from the 1930s discovery of remnants of chlorophyll in oil deposits by the German chemist Alfred Treibs . At first, the theory concluded that oil was formed from plant matter on land that contains so much of the land-based chlorophyll and shallow sea-based plants. Later, scientists discovered that offshore oil deposits also contained molecules typically associated with microbes that survive on the ocean floor. These microscopic organisms are so small that a single drop of seawater may contain a million such organisms.

Now, modern science can be used to associate different microbes to different grades of oil. We can then deduce that the precursors to oil will be found where there are prolific swamps or seas of algae during periods of high atmospheric carbon dioxide concentration. The various ages of these microbe fossils give oil geologists information about the length of time the process of heat and pressure has been taking to act on an oil deposit. The unique combination of organisms, heat, pressure, and time gives each oil field its distinct characteristics.

The combination of heat and pressure then converts these biomasses into various mixes of hydrocarbons, based on their particular combination of material and depth.

This optimum combination of light and heat, carbon dioxide and water, can create an environment in which these microbes and forms of biomass grow and die more rapidly than the ability of the seabed or swamp floor to decay and redistribute them. When this occurs, layers of biomass get compounded in a form of ever-thickening black bio-mud. The theory explains why, after more than a hundred years of exploration and extraction, we have found oil in areas of land that had once been undersea, and why, as those sources of easier-to-find oil are exhausted, we find ourselves increasingly drilling offshore and in deeper waters.

Another factor that can accelerate the oil-making process is the abundant inflow of nutrients from rivers and streams. We now often find oil in gulfs where big rivers meet the ocean. This constant feeding of nutrients caused carbon-based organisms to amass and, subsequently, to be buried as the region is inundated with sand and silt layers. Time and pressure caused the sand, clay, and silt to turn to sandstone, entombing the primordial organic mud. Cut off from an oxygen-rich environment, and buried deeper and deeper by insulating sand and stone forming above, the organic mud cannot decay and dissipate. Instead, the mud rose in temperature from the ambient heat of the earth's core, and was cooked and baked in a slow chemical process that consumes any remaining oxygen and produces hydrocarbons. Where the organic mud can cook at a temperature somewhere between hot and boiling water, the earth created the ideal environment for the production of oil.

We now see that this optimal environment existed at the confluence of the Mississippi River in the Gulf of Mexico, the region in South America fed by the Amazon basin, and the area in the Persian Gulf and Middle East that once contained ancient oceans.

Following this theory to its natural conclusion, we can surmise that future oil will be found increasingly in deeper offshore sites. Cambridge Energy Associates observe that deepwater hydrocarbon extraction, defined as deeper than 2,000 feet, has more than tripled in the past decade, and is projected to double again by 2015.[5] They also find that deepwater discoveries now make up the majority of all discoveries, and represent significantly larger fields. Increasingly, deepwater exploration and extraction are driving higher U.S. oil production. These new discoveries have allowed the United States to demonstrate greater year-to-year production for the first time in almost two decades.

It has been estimated that humankind has used about one trillion barrels of oil. In 2009, the *Oil and Gas Journal* estimated that world crude oil reserves represented another 1.342 billion barrels, with a further 6,254 trillion cubic feet, or 1.08 billion barrels oil equivalent (BOE), of natural gas.[6] Of course, the estimate of remaining reserves depends on both the price we are willing to pay and the success of new technologies to extract the remaining oil. In any event, experts believe that conventional oil reserves will last for another 40 years at current consumption levels. This time horizon may be shortened significantly if our pattern of economic growth and demographics shifts toward the rapidly growing economies of China and India.

Interestingly, in the recent human era in which we have been relying on the burning of these oils, we are once again returning the sequestered carbon dioxide to the atmosphere, perhaps to give rise to new oil reserves millions of years from now. Obviously, by then, our planet will have moved well beyond the use of oil as the primary energy source.

4
The Science and Refining of Oil

Hydrocarbons literally fuel our economy. Oil is a major constituent of the class of energy labeled nonrenewables that also includes natural gas and coal. The other sources of energy are nuclear, which technically is also a nonrenewable, and renewable energy sources, ranging from wind and solar power, biomass, wave power, and hydroelectric. With the exception of nuclear energy, these energy sources derive their energy content from the sun, through differential heating of the atmosphere, the photons emitted from the Sun and impinging on solar panels, heating fluids, from the creation of weather, or from the absorption of light and carbon dioxide to create carbohydrates in plants.

Oil remains the single largest energy source for the world, followed by coal, natural gas, nuclear power, and hydroelectricity.

The global energy consumption (shown in Figure 4.1) has been growing at a consistent rate of 1.8% per year, in line with average global economic growth. While in 2009, energy consumption declined for the first time, in response to the global economic meltdown, oil has also been the fastest growing component of energy consumption over the past quarter century.

Oil has been consistently the largest single component of energy usage, for a variety of reasons. The vast majority of energy is used for either heating or for the powering of electrical devices. Even engines in our vehicles are heat engines. They use the heat of burning fuel to expand air and push a piston against a load to generate motion or power. Electric motors can more efficiently produce motion without the waste heat engines discard in exhaust. However, apart from hydroelectric electricity, the generation of electricity also often relies on heat engines to turn generators.

While hydroelectricity is very efficient, it depends on the right physical conditions to trap large amounts of water in constantly replenished

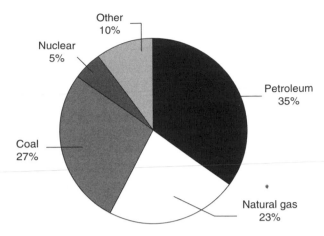

Figure 4.1   Global energy consumption derived from U.S. Energy Information Administration data

Source: http://www.eia.doe.gov/oiaf/ieo/ieorefcase.html, table A2, accessed October 22, 2010.

reservoirs. These sites take up huge amounts of land, and hence are located in isolated regions usually far away from the populations they serve. Like oil, the energy must then be transported to market. The cost of electricity transportation is relatively high, often consuming 10%–20% of the electrical energy in the process.

On the other hand, a primary advantage of oil is that it is easy to transport. For instance, supertankers transport oil around the world for pennies per gallon. The energy loss in transportation is very small, making oil ubiquitous worldwide. Oil also contains a large amount of recoverable energy per unit volume or weight, further contributing to its transportation efficiency.

While we speak monolithically about oil, we actually consume oil in a number of ways. More correctly, we should refer to hydrocarbon consumption, and differentiate between the various types of hydrocarbons produced by oil and consumed for their different energy contents.

Various types of hydrocarbons

As the name implies, hydrocarbons are molecules that combine the two elements hydrogen and carbon. Carbon is the basic building block of all living organisms, and is the element that defines organic chemistry. The saturated hydrocarbons that are constituted primarily of combinations of carbon and hydrogen include paraffins, alkanes, and cycloalkanes, the alcohols that combine carbon with oxygen and hydrogen in the form of a hydroxyl (OH) group, and carbohydrates that combine carbon with hydroxyls to produce sugars. All are variations of the simplest molecules that provide chemical fuel to much of what makes up our environment.

For instance, our body consumes carbohydrates, mostly in the form of sugars, to produce energy in our muscles. The energy is used, and the sugars are transformed into carbon dioxide and water. Likewise, alcohol can be burned in an oxygen-rich environment to produce energy, water, and carbon dioxide.

The simplest alcohol is methanol, with a chemical formula CH_3OH. When two molecules of methanol are ignited by combining them with three molecules of oxygen O_2, energy is given off, in addition to two molecules of carbon dioxide CO_2 and four molecules of water H_2O. We can reverse this process by combining water, carbon dioxide, and energy to produce methanol. We are familiar with natural organisms called yeasts that perform this process to create the alcohol in wine and beer. Likewise, chlorophyll takes the energy from sunlight and carbon dioxide from the atmosphere to create sugars for plants and oxygen that allows animals to survive.

The hydrocarbon combinations that are based solely on carbon and hydrogen provide energy when the carbon-hydrogen bond in the molecule is oxidized, or severed, resulting in the formation of smaller molecules of carbon dioxide and water. Hydrocarbons function very much like the alcohols and sugars. For instances, paraffins combine carbon atoms solely with hydrogen atoms in a ratio C_nH_{2n+2}. The slightly lighter-than-air-gas methane, which is the primary constituent of natural gas, combines one atom of carbon with four of hydrogen. The formula for this paraffin, also called an alkane, labelled C1, is CH_4. Similarly, the next larger paraffin ethane, C_2H_5, is a heavier-than-air-gas that, with methane, makes up natural gas. Propane, also heavier than air, is labeled C3, and has the formula C_3H_8. The heaviest paraffin that is still a gas at room temperature is the C4 we know as butane, C_4H_{10}.

Heavier combinations of hydrogen and carbon, from pentane C5 to C17, remain a liquid at room temperature. Still heavier paraffins run from C18 and above, and remain a solid at room temperature.

Hydrocarbons are classified based on the range of carbon atoms in the various molecules. The heaviest, paraffin waxes, fall in the range of 20–40 carbon atoms per hydrocarbon molecule. However, the term "paraffin" can refer to any linear, or normal, hydrocarbon in which the carbon atoms are linked to each other in a chain using a single bond.

Because carbon permits four bonds, there remain three other bonds for associated hydrogen atoms. These paraffins represent about a third of the weight of crude oil.

Alkenes are related to paraffins except that each of the carbon atoms is double-bonded to another, resulting in fewer remaining bonds for hydrogen atoms. Some commonly found alkenes include ethylene, butene, and isobutene.

Half of crude oil by weight, on average, is made up of naphthenes. These hydrogen-saturated carbon atoms are made up of one or more rings of carbon atoms, with the remaining bonds saturated by hydrogen in a ratio of $C_nH_{2(n+1-g)}$, where g is the number of carbon rings. The ring nature of the naphthenes are differentiated from the linear alkanes through the prefix cyclo-, that is, cyclopropane, cyclobutane, cyclopentane, and so on.

Crude oil is also made up of aromatics and asphaltics, in smaller amounts. Aromatics are similar to naphthenes, but with single and double bonds in their rings of carbon, with the remaining bonds joined to surrounding hydrogen atoms. Finally, asphaltics are molecules of carbon, hydrogen, oxygen, nitrogen, and sulfur that can remain once the hydrocarbon molecules in crude oil are distilled off in the refining process. The remaining materials can be used to produce asphalt, the tar that is used for roadways.

Crude oil differs in its composition of these various components of alkanes and alkenes, naphthenes, aromatics, and asphaltics. An oil that has a larger mix of low carbon number alkenes is called "light," while heavy crude contains a greater proportion of the larger hydrocarbons. Sweet crude contains little sulfur, while sour crude may contain 6% sulfur or more. The famous West Texas light sweet crude oil was ideal for gasoline production because it contained little sulfur and a higher share of the lighter, more volatile hydrocarbons needed to produce the gasoline that is sufficiently spontaneous and explosive to function in a spark ignition engine.

While the natural gas that fuels many homes, stoves, and power plants is predominantly methane, most of the other hydrocarbons we consume are a combination of various alkanes. For instance, hexane C6 through decane C10, in specific ratios, produces gasoline, diesel fuel, and aviation fuel. The less viscous gasoline relies on greater proportions of the lighter alkanes in the range, while increasingly thick and viscous hydrocarbons are made up with greater proportions of heavier alkanes.

The amount of energy contained in the various distillates of crude oil are related to the number of conversions of strong CH_2 bonds to less robust bonds in the resulting carbon dioxide CO_2 and water H_2O. For instance, with the cycloalkanes that make up the bulk of crude oil, the amount of energy that can be released is almost directly proportional to the number of CH_2 carbon-hydrogen combinations in the various hydrocarbon molecules (Figure 4.2) where kcal is the number of kilocalories of energy released in combustion, and a mole is a measure of the number of molecules consumed in combustion.[7] While the larger hydrocarbons can release proportionally more energy, the energy per unit of weight of hydrocarbon remains relatively constant. Hence, the energy content of hydrocarbons is differentiated by the density, or weight per liter or gallon, of the fuel.

The process of refining oil into its constituent components of alkenes and cycloalkanes relies on the fact that the lighter hydrocarbons are less dense, and hence are gaseous at lower temperatures. The traditional method to separate crude oil into these constituent molecules is through fractional distillation. When heat is gently applied to petroleum, the lightest molecules turn to a gaseous state first. The lighter molecules are then removed and are progressively allowed to condense back to their pure liquid forms. As the temperature of the petroleum rises, each molecule is distilled off until all of the constituent liquids have been separated and only the tars remain.

Modern refining of crude oil takes advantage of other less expensive or more efficient processes to break some of the molecules up into smaller molecules. For instance, catalytic cracking uses metals or other catalysts to promote the breaking of some larger hydrocarbons into smaller ones that can be used to make more volatile products such as gasoline that commands a higher price in the market. While these catalytic reactions

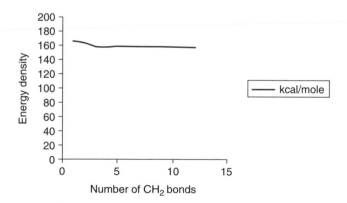

Figure 4.2   Numbering of alkanes and energy equivalents

still require some energy, the nature of a catalytic reaction leaves the catalyst unchanged to efficiently permit further identical cracking reactions.

Once refined into its constituent parts, these liquids can be recombined to produce mixtures of fuel with the desired properties of volatility, density, and energy content. For instance, a 42 gallon barrel of light sweet crude oil can produce about 19 gallons of gasoline, made up primarily of C4 to C12. Other, heavier crudes are more amenable to the production of diesel or fuel oil, bunker oil, or other products that are heavier, denser, and less volatile.

While the science of petroleum refining is well understood, it is not without risks. The combination of heat, reactive catalysts, chemicals used for chemical cracking, and the proximity of highly volatile and explosive products cannot be without hazard. Despite safety precautions, human error, defective valves, pipes corroded through constant contact with caustic chemicals and volatile hydrocarbons, and environmental factors create significant risks that must be managed but can never be reduced completely.

While the focus of this case study is the inherent risk and environmental cost of crude oil exploration, drilling, extraction, and transportation, we must remain aware of other dangers associated with a hydrocarbon-based economy. An oil-based economy must manage the inherent risk created by volatile hydrocarbons, from refining and transportation to market, to refueling, highway risks, and the environmental consequences of burning hydrocarbons. All these factors contribute to the risks of an economy that derives a majority of its energy from volatile and highly combustible materials created millions of years ago and safely contained under land and under seas, until now.

5
Oil in Our Past and Present

While the natural history of oil punctuates the earth's natural history, its role as the primary energy source for humanity spans little more than a century.

Petroleum has been used by humans for millennia. The petroleum that occurred naturally along the shores of the Euphrates River was used for medicinal purposes, while naturally exposed tar was used as a building material 4000 years ago in the towers of Babylon. However, it was not until the 1850s that the distillation process was discovered to refine kerosene from crude oil for lamps. This process allowed kerosene to replace the increasingly dear whale oil previously used.[8]

While the first wells were produced in Europe and West Asia, soon oil was discovered in the U.S. states of Pennsylvania and Texas. This newfound oil fueled the rapid expansion of the U.S. economy in its Gilded Age, and contributed to the U.S. overtaking Great Britain as the world's largest economy in the early part of the 20th century.

Indeed, Rockefeller's Standard Oil Company became the notorious vertically integrated monopoly that would eventually define corporate excess and create the clash of large companies and government policy bent on curtailing their strength. Standard Oil's efforts to monopolize the entire oil and fuel industry led to a showdown with the U.S. Congress and the definition of modern antitrust policy. Ultimately, Standard Oil (S.O.) was broken up by government fiat, replaced, in part, by a namesake Esso that would eventually become Exxon and, in a merger with Mobil Oil, ExxonMobil. Notwithstanding attempts to limit the influence of U.S. oil companies, these companies dominated global oil production by the middle of the 20th century, and retained their global energy production dominance until Saudi Arabia and Russia each surpassed U.S. production in the 1970s.

Now, the Middle East is the dominant region for economic oil reserves, while Canada's unconventional oil reserves, in the form of hydrocarbon-saturated bitumen and tar sands, leads the world in reserves when the price of oil rises sufficiently to make their higher extraction costs economical.

The price of oil

The price of oil is determined by the interplay between market demand and the scarcity of oil. We first look at the factors that influence demand, and then explore the nature of scarcity and supply.

The demand for oil can be represented on a graph as a line that defines the relationship between the price of oil, on the vertical axis, and the quantity demanded on the horizontal axis. The law of demand states that a consumer demands less of a good as the price of the good increases, and vice versa. This negative relationship between price and quantity demanded is shown through the downward sloping demand curve in Figure 5.1.

A nation demands oil based on the value of its consumption and mindful of the price of other alternative energy sources. For commodities such as oil, demand rises rapidly as household and national income rises. Such commodities, called luxury goods, represent a much larger share of purchases for the more developed nations. For example, residents of a high income nation will heat larger houses, drive more and larger vehicles, and consume more energy, much of which is hydrocarbon fueled.

The price of alternatives to oil also influences its demand. A rising price of oil will make economical alternatives that were subeconomic

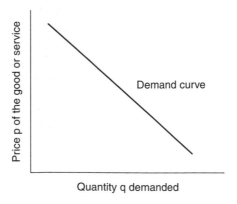

Figure 5.1   The graph of a demand curve

previously, wood or electricity, for instance. These alternatives are called substitutes because we replace other energy sources for oil consumption as oil prices rise.

Other goods, called complements, may also affect the price of oil. For instance, if the price of automobiles falls or their demand increases, the demand for oil likewise increases, and draws up its price.

Our demand for oil also depends on our income, population growth, and socio-economic norms. A higher national income increases demand for oil through higher personal incomes. Such a growing gross domestic product is often associated with greater industrialization, and hence greater energy use intensity. A growing population likewise corresponds to a greater number of households and a greater need for automobiles and heating furnaces. And, socio-economic expectations that shift toward greater automobile use or energy intensity will increase the demand for oil. All of these factors combine to create an increased level of energy consumption as a nation's gross domestic production rises. For instance, we see that the developed countries that are members of the Organization for Economic Cooperation and Development (OECD) have relatively modest expected energy growth, while the emerging nations of the world show more dramatic increases in energy demand (measured in quadrillion British Thermal Units, or qbtu) as their economies grow (Figure 5.2).

These factors can be used to predict the future price and consumption of a commodity such as oil (as we will see in the next chapter). For instance, speculators must model the change in demand that would result as oil's substitutes and complements change in price, income for nations and for households change, and demographics and population evolve over time. The effects of these patterns is modeled to predict the

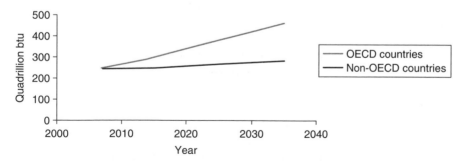

Figure 5.2 Worldwide energy demand expectations by the Energy Information Agency of the United States Government
Source: http://www.eia.doe.gov/oiaf/ieo/highlights.html, accessed October 22, 2010.

percentage change in the quantity of oil demand based on a percentage change in these other factors, known as "elasticity." We can also measure the percentage change of the price of oil that arises from a percentage change in the quantity demanded. It is this change in the price of crude oil that acts as the incentive for oil exploration companies to drill deeper through rock, deeper in the ocean, and at greater risks.

For a commodity such as oil, we find that the quantity demanded does not respond much to a change in its price. This insensitivity, or inelasticity, arises because, in the short run, most people will not travel significantly less if the price of gasoline rises, at least until it becomes more economical to buy a fuel efficient vehicle. In the long run, however, prolonged high oil prices will induce people to become more energy efficient, and will induce suppliers to develop alternatives to oil. With a greater number of economical substitutes to oil, the demand curve for oil becomes flatter as the quantity of oil demanded becomes more sensitive, or elastic, to its price.

Producers find commodities with such inelasticity profitable because consumers will not demand appreciably less as the price increases. Traditionally, a supplier can attempt to enhance consumer loyalty through advertising or by absorbing its competitors. However, commodities such as oil or gasoline are neither prone to consumer loyalty nor heavily influenced by advertising. Monopolization is the most effective method of driving up oil and gasoline prices through strategic reductions in supply.

The supply side

The unique quality of oil and other depletable resources is that its long-run supply is relatively fixed. Unlike other goods or commodities for which we can simply make more, our only response with greater demand for oil is to find new, more expensive, and harder-to-reach energy sources. At some point, the resulting price will rise so significantly that we will be forced to transition to the next most economical alternative. Such a transition point is likely still decades of price increases away.

In the same way that we modeled demand, the price of a commodity is related to the quantity supplied and can be drawn on a graph called the supply curve. For oil, a rising price makes previously marginal, or subeconomic, fields profitable. The supply curve then traces out the mix of oil wells ranked by the breakeven extraction price of the oil that lies beneath. Oil from wells that can be pumped and brought to market at a price lower than the prevailing market price yields a high profit. Oil from wells that can barely be brought to market affordably at the prevailing price receives little profit. A supply curve then slopes upward, indicating that a higher price makes more wells profitable and creates a higher potential supply of oil that can be brought profitably to market.

The intersection between demand and supply determines both the equilibrium price of oil and the quantity demanded and supplied. For instance, with a price of oil that ranged between $70 and $80 per barrel of crude in 2010 resulted in demand in the range of 60 million barrels consumed worldwide per day.

However, commodities such as oil for which supply cannot be expanded significantly in the short run, and no faster than the pace of exploration and advancing technology in the long run, cannot expand quickly either to a spike in demand or to a spike in speculative interest. Correspondingly, small increases in expected demand can result in large increases in the price of oil at world auction. Likewise, economic downturns can result in plunging prices.

As we have described earlier, oil is a commodity in decidedly fixed supply. Much of the oil we consume today was produced in the Paleozoic Era more than three hundred million years ago as leafy plants, prehistoric forests, and sea-based algae grew and then died in such a carbon dioxide-rich environment that it overwhelmed the environment's ability to absorb it. As this organic material was layered from above on the bottom of seas and swamps by sand which turned into sandstone and rock, the temperature and pressure baked the organic materials into what we consume today to heat our homes and run our vehicles.

An estimated trillion barrels of oil already discovered and consumed. More than another trillion barrels of proven reserves can be extracted

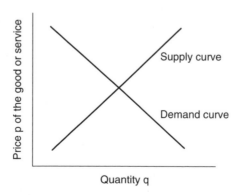

Figure 5.3 The supply and demand for a commodity

at current prices. There is yet another trillion barrels that could be economically feasible if the price of oil was to rise sufficiently, in tar sands near the surface in parts of Canada, and buried deeper in Venezuela. However, the Second Economic World nations are increasing their rate of usage of vehicles and their rate of manufacturing as they aspire to imitate the energy-intensive patterns of the First Economic World. We will look at these expected patterns of increased demand in the coming chapter. For now, though, we take a look at options that will act as a backstop technology to which we will transition as oil runs out.

Backstop technologies

The dramatic growth in the pace of industrialization in Brazil, Russia, India, and China (the BRIC nations) has driven oil prices up to the point that there began to be considerable interest in alternative energy technologies.

Certainly, the lowest hanging fruit is for greater conservation and enhanced efficiency of the energy we have. Better insulation, more fuel economic hybrid gasoline-electric automobiles, a greater use of diesel technology, and lighter vehicles allowed the same amount of fuel to go further.

Ethanol from corn and biomass, or vegetable-based oil from plants and algae can also substitute for crude oil, but still at a significantly higher price.

In addition, new technologies that allow vehicles to be run on battery power permits an economy to use oil more efficiently. This efficiency arises when electricity is generated in large oil, gas, or coal fired power plants. In turn, the net rate of oil consumption falls.

Oil consumption may also decrease if other technologies are used to generate electricity. Wind, solar, and nuclear technologies can be used to provide some, though not a sizeable portion of energy for transportation.

An energy renaissance

Natural gas is often used as a substitute for oil. It can be an economical substitute for home heating, and can power oil-fired electric generation plants after some modifications. Accordingly, it is priced relative to the price of oil.

Reserves of natural gas actually exceed the amount of known oil reserves. However, while natural gas is abundant and is cleaner burning than oil, it is still a greenhouse gas producer. It is also more difficult to transport and does not offer the driving range for vehicles converted to operate on natural gas.

These factors combine to limit the application of natural gas primarily to those areas which have some regional source, or to those areas that can be linked to a gas supply through a natural gas pipeline distribution network.

Coal, too, may offer a source of energy that could last for decades, but not indefinitely. However, even as aspiring nations increasingly rely on abundant coal to fuel their growth, our growing awareness of the problem of global warming make carbon-rich coal burning an unattractive prospect.

All nuclear power relies on the insight Albert Einstein first realized. While mass and energy, in total, cannot be created or destroyed, we can change mass into energy, and vice versa. Einstein's most famous equation, $E = mc^2$, introduced to the world in his Special Theory of Relativity, taught us that we can create energy from mass. The process of nuclear fusion used this equivalency in the Big Bang, and ever since, to fuel our stars and fuse together the elements that make up everything around us. Scientists learned in the 1940s and 1950s to imitate uncontrolled fusion in the hydrogen bomb (H-bomb).

While scientists have been trying for decades to replicate this process in a controlled way that could provide our economy with limitless and abundant electrical energy and heat, most agree that such energy elusively remains generations away.

The Big Bang also left radioactive uranium and thorium underground that slowly degrades, and gives off heat, through nuclear fission. If these materials are mined and their degradation accelerated, we could generate prodigious amounts of heat today. This heat could be used to make steam to turn turbines and power electric generators.

New nuclear plant designs in Europe are now employing third-generation technologies and are formulating fourth-generation designs. The antiquated first- and second-generation nuclear power plants developed in the 1950s that first harnessed the heat of fission can now be replaced by newer third- and fourth-generation plants that can generate electric power by consuming the nuclear waste left untapped, slowly degrading, and stored at old nuclear power plants. These new plants can convert waste with half-lives of millennia into materials that are much safer and with half-lives much shorter. At the same time, these newer plants run 20 times more efficiently than the plants they could clean up and then render obsolete.

Geothermal energy also holds some potential for areas that are prone to geysers or volcanoes. The high heat below the surface of the earth can be used to run turbines and electric generators, and can heat build-ings or entire towns. Like nuclear energy, it uses an abundant energy source below the earth's surface that is technically nonrenewable.

Renewable energy sources

The various forms of renewable energy all rely on one common ele-ment – the energy of the sun impinging on the earth.

This energy can be most obviously and directly harnessed in photo-voltaic cells that directly convert light energy to electric energy. The current state of these cells allow for a conversion efficiency in the order of 5%–30%. While the technologies remain costly, and can only gener-ate significant energy during the day and in high solar radiation areas, costs are coming down rapidly. It is likely that solar energy will move beyond the niche market for those who do not have access to cheaper alternatives to the more mainstream energy market as costs continue to decline.

Wind energy, too, is technically derived from solar power. The heat of the sun varies by latitude, and creates pockets of differing ambient air temperature and pressure. Equalization of these pressures across the earth, when combined with the rotation of the earth, creates the prevailing winds that power modern wind generators. These giant energy generat-ing mills can stand 400 feet tall, and can generate upwards of three mega-watts of energy. One such wind-powered generator in a suitable wind-rich location can generate enough electric power to take care of the electric needs of almost 3,000 homes. At a cost upward of $3 million to $5 mil-lion per wind generator, the technology remains expensive. However, it is becoming more economical for some locations that have sufficient wind resources and are not too far from an existing transmission line that would allow the power to merge with the electric energy grid.

Finally, biomass for the conversion to energy is an exciting energy prospect. Ethanol, derived from corn fermentation in the United States, and sugar beets, in Brazil, have been a substitute for gasoline, either as a mix with gasoline or as a sole fuel for specially equipped vehicles. This technology converts carbon dioxide to sugars, which are then con-verted to ethanol fuel. Consequently, these biomass fuels are carbon neutral. They absorb as much carbon dioxide in their biomass growth as they release in their use as a fuel.

However, some note that this energy equation is not balanced if one considers the amount of petroleum used to make and transport the fertilizers necessary to grow corn, to plant and harvest corn, or to fer-ment and then distill the ethanol. The intensive use of corn for ethanol may also displace other crops, necessitate increased transportation and farming resources, and thus increase fuel usage, and raise the cost of food. Clearing of land, which had also been acting as carbon sinks to make way for new corn fields, may also tip the energy equation away from ethanol production from food crops.

Techniques are being developed to produce ethanol and vegetable oils from other more benign biomass sources. Plants are being developed that can convert wood, waste wood chips, and grasses to ethanol. Varieties of algae are being genetically engineered to absorb the energy of the sun to create oils that can substitute for diesel fuel and kerosene. Some of these fuels, from algae and from beans and oil-producing plants, have even been approved to fuel commercial jet airplanes.

The hydrogen economy

There is much discussion about a hydrogen-based economy. The use of hydrogen as a fuel has some advantages over oil. Hydrogen can offer similar energy densities as we have grown accustomed to from oil. It burns very clean and can be converted directly to electricity using fuel cell technologies. It does not produce any carbon dioxide in its consumption, and hence is viewed as more environment-friendly. With some advancements in storage technologies, hydrogen may soon be able to offer us a range in our vehicles and a rapid refueling that makes long distance transportation efficient and economical.

However, hydrogen is not an energy source in itself. As a lighter-than-air gas, any natural deposits have long since disappeared. Instead, hydrogen is created through abundant access to electricity or very high heat, as can be produced in modern nuclear power plants. Because the creation of hydrogen requires another energy source, it should instead be viewed as an energy storage technology rather than as an energy source.

These various alternatives to oil will certainly make up a larger share of our energy portfolio in the future. However, they will do so only as the price of oil rises with increased scarcity. They will not immediately replace oil. Rather, they will allow existing reserves of oil to last longer.

While we may have a relatively good understanding of the supply of oil and the feasibility of various alternatives to oil, the wild card is in the future demand for oil. It is this subject we turn to next.

6
Demand for Oil in Our Future

There is little doubt that oil demand will increase significantly as aspiring BRIC nations (Brazil, Russia, India, and China) join the club in the First Economic World. Mitigating this rise in demand will be the ability of the most developed nations to afford new and more efficient technologies. However, if hand-me-down technologies make their way from the First Economic World to newly industrializing nations, and as these nations demand the commodity that they could not at one time afford, oil demand will still increase significantly. We next look at these global economic development phenomena through shifts in worldwide population and wealth.

To best understand the inevitability of an increase in global energy consumption, we must explore the relative growth of the Second and Third Economic World economies as they develop and converge to join the First Economic World. We must then compare their relative populations and extrapolate the implications to global energy demand.

The development cycle

The industrial revolution brought regular and spectacular double digit real economic growth, in waves, to the countries in the First Economic World that now constitute the membership of the OECD.[9] An economist named Simon Kuznets demonstrated this phenomenon. He noted that the establishment of property rights and the creation of an active entrepreneurial class initially generate economic growth upwards of 10% to 20% annually. This rapid growth of energy demand comes at the expense of unsustainable consumption and of income inequality. Nonetheless, a rapidly growing middle class eventually transitions an aspiring nation to one of more service-oriented production, the growth

30

of government, suburbanization, and a higher, but slower growing rate of energy consumption.

While the Cold War categorized the First World nations as aligned with the United States and its allies, the Second World with the former Soviet Union, and the Third World with those unaligned with either, it makes more sense now to group nations based on their economic status. We can define those First Economic World nations as the countries that are economically developed and reside to the right of the curve in Figure 6.1 below. Meanwhile, the Second Economic World nations are developing rapidly and are often experiencing double digit economic growth. Finally, countries of the Third Economic World do not have the well-established property rights, the good government, or the level of education and investment necessary to industrialize rapidly.

The emerging middle class demands greater health and safety standards, higher education, a greater regard for the environment, and more luxury goods, including automobiles and energy. Eventually, this demand levels off, as does energy consumption. When growth levels off at a more sustainable rate of 2%–4%, maintained primarily through technological and productivity improvements, and a reduced rate of family formation, we find we have fallen off to the right of the growth path shown in Figure 6.1.

This consistently observed phenomenon of spectacular growth in a rapidly developing nation, followed by even and more sustainable growth when the economy matures, is an inevitable conclusion of the Law of Diminishing Returns. A country can develop rapidly as it picks the low hanging economic fruit of new industrialization, urbanization,

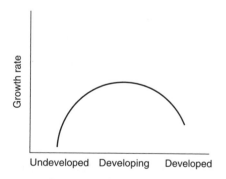

Figure 6.1   Rate of growth of GDP and energy demand over the development cycle

---

and education. Once the economy is industrialized, urbanized, and educated, its rate of family formation drops, its growth becomes more dependent on a service sector that grows less spectacularly, and it finds itself squarely in the First Economic World.

The role of emerging demographics

The rate of industrialization is only part of the story. We also find that increased wealth creates a reduced need for family formation, and a dramatically changing population demographic.

The Third Economic World has a high gross rate of family formation, but a low net rate, as disease and famine stymie population growth. These nations find it difficult to reach a critical mass of education, wealth creation, industrialization, and good government. Until they overcome these hurdles, growth cannot progress.

Once a country is able to shed the shackles of hampered economic growth, a Third Economic World nation can transition into the rapid growth of an aspiring nation in the Second Economic World. Finally, nations that experience their own Gilded Age and settle into the characteristic pattern of education and urbanization of a developed nation become members of the First Economic World.

We can most easily see the implications of this phenomenon by comparing relative populations and projections as published by the United Nations Population Division. In a recent report, they predict the patterns of growth worldwide as shown in Table 6.1.

These results are most powerfully demonstrated in Figure 6.2 which tracks expected population growth to the year 2150.

Table 6.1 World Population 1750–2150 (millions) as projected by the United Nations Population Division

	Year							
Region	1750	1800	1850	1900	1950	1999	2050	2150
North America, Europe, Oceania	167	212	314	496	732	1,066	1,066	966
Asia, South and Central America	518	659	847	1,021	1,569	4,145	6,077	6,473
Africa	106	107	111	133	220	767	1,766	2,307
World	791	978	1,262	1,650	2,521	5,978	8,909	9,746

Source: Read, Colin, "The Global Financial Meltdown," Palgrave MacMillan Press, London, 2008.

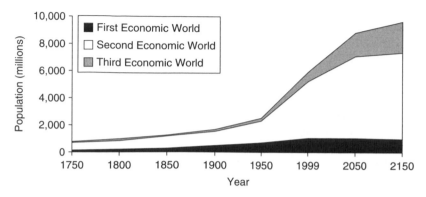

Figure 6.2   Population growth in three economic worlds

The graph demonstrates that those nations in the First Economic World grew rather rapidly, but has begun to peak out. Meanwhile, the BRIC nations and others in the Second Economic World are growing very rapidly, and will continue to grow over for another century. These Second Economic World nations already constitute the bulk of the world's population. Growth beyond 2150 may then be sustained as the Third Economic World could assume the mantle of population growth, likely followed by economic growth.

Before we tally the effects on oil consumption and prices based on economic growth, we should note that the Second Economic World is rapidly converging with the First Economic World.

China surpassed Japan as the world's second largest economy in 2010. If one extrapolates the rapid average growth in China from 1999 to 2009, compared with the more subdued growth in the United States, the world's largest economy over the same period, we find that China will surpass the United States as the world's largest economy by 2021.

Similarly, the nations of Asia and Central and South America will join North America as affluent, energy-intensive consumers. Meanwhile, these nations will look toward Africa for resources, which will hasten their transition to Second Economic World economies.

Finally, we should note that these converging nations will also urbanize as had their First Economic World counterparts. Indeed, it is this potential for urbanization that acts as the greatest engine for economic development. Data from the 2003 Revision of the World Urbanization Prospects, created by the United Nations Population Division, shows the rapid urbanization, coincident with the pace of economic development (Table 6.2).

Table 6.2 Urbanization and global population 1950–2030 (millions)

Region	Year				
	1950	1975	2000	2003	2030
Northern America, Europe, Oceania – Urban	398	641	802	815	930
Asia, South and Central America – Urban	302	772	1760	1900	3266
Africa – Urban	33	103	295	329	748
North America, Europe, Oceania – Rural	334	300	273	270	203
Asia, South and Central America – Rural	1263	1948	2440	2467	2331
Africa – Rural	188	305	500	521	650

Source: as projected by the United Nations Population Division.

The emerging markets of the Second Economic World do not yet consume oil nearly at the rate of the First Economic World. However, these developing nations, with the developed nations will be fully integrated into the First Economic World within the century. Consequently, it is reasonable to predict that, with convergence, oil demand will continue to accelerate until at least 2050, and likely until the end of the century.

When these Second Economic World nations fully converge into the service-intensive developed nations, their energy-intensive manufacturing sectors will transition toward currently undeveloped nations. Nonetheless, resource usage, especially energy usage, will continue to grow until it plateaus at the higher levels associated with the First Economic World – unless we are able to transition to new and sustainable energy sources.

Peak oil

Another way to view convergence is to compare energy consumption of the Organization for Economic Cooperation and Development (OECD). Its members are Australia, Austria, Belgium, Canada, Chile, Czech Republic, Denmark, Finland, France, Germany, Greece, Hungary, Iceland, Ireland, Israel, Italy, Japan, Korea, Luxembourg, Mexico, the Netherlands, New Zealand, Norway, Poland, Portugal, Slovak Republic, Slovenia, Spain, Sweden, Switzerland, Turkey, United Kingdom, and the United States.

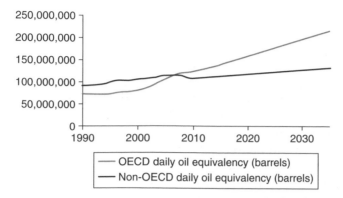

Figure 6.3 Marketed daily energy consumption in barrels of oil equivalency

The U.S. Energy Information Agency estimates energy consumption among the OECD First Economic World and the rest of the world's nations in Figure 6.3.[10]

As the data shows, increased agriculture, urbanization, and manufacturing all compound into greater economic growth, and substantially greater oil consumption by the year 2035. From the year 2000 to the year 2035, global energy consumption, primarily in the form of oil, will have nearly doubled. However, our previous analysis demonstrated that global economic growth and population is expected to accelerate well beyond 2035.

Another way to extrapolate oil consumption is in the annual average rate of production per capita. For the past three decades, this consumption has remained within the range of between 4.44 and 5.26 barrels per person per year, based on the current pattern of industrialization. The United States consumes just under 25 barrels per person per year, or approximately five times the global rate. The global population is expected to increase by almost 40% from the year 2000 to 2050. Convergence may result in oil demand that grows significantly more rapidly, depending on the rate that the Second Economic World replicates the high energy consumption patterns of the First Economic World.

The dynamic of oil prices will depend on the level by which the increased demand outstrips the ability of the expansion of economic oil reserves to keep up. If demand outstrips supply, and if alternatives to oil remain expensive, the price premium on oil will expand. If so, there will be even greater pressure to drill in higher risk areas. However, if alternatives to oil are more rapidly developed, this pressure is reduced substantially as we transition earlier than expected to sources that can offer a long-term and sustainable alternative to oil.

The greatest single factor that will influence the life of oil is the creation of viable substitutes. If research and development is insufficient to signal an economical substitute within the next decade or so, we will begin to see oil prices spike and oil exploration take even more risks. The price of sustainable energy will dictate the price and risk of oil exploration and production.

The challenge will be to discover sustainable energy sources that can be replicated at a sufficient scale to substitute for 10–20 million barrels of oil equivalent per day. We see that battery technologies may outstrip our ability to supply lithium and zinc. We know that it is difficult to find wind resources that can efficiently meet our needs. While solar radiation is abundant, the cost of these materials remains prohibitive. And, while we can create biomass to provide an alternative source of liquid fuels, the cost is the land and the crops that must be diverted away from other human needs toward the satisfaction of our energy needs.

Unless these resource constraints can be overcome in ways not yet understood, an increasing price of oil, and increasing pressures to discover and produce more oil, seems inevitable.

7
The Industry of Oil Extraction

The intensive use of petroleum-based oil is a relatively recent economic phenomenon. While the term "oil" is derived from the Greek term "elaion,", for olive oil, the term can be used for any material that is liquid at room temperature, will not mix with water, and is soluble by organic solvents. Oils are a combination of carbon and hydrogen, and can refer to either vegetable-based or petroleum-based oils. It is the search for petroleum that now induces us to go to lengths unimaginable just a few decades ago.

In essence, all oils originate from organic processes. No oil is produced solely through geological processes in the absence of organisms. Even the mineral oils we most often associate with crude oil have their origin in the decomposition of organic matter.

The creation of crude oil requires heat to cook, and rock or depth to pressurize primordial masses of decaying organic matter. These two essential factors of heat and pressure dictate that oil will be formed beneath layers of inorganic matter, and baked at subterranean temperatures.

Discovery of oil is most likely when this process occurs relatively close to the surface, in areas that support the human populations who have learned to use crude oil. Before the industrial revolution, this easy-to-access oil was found seeping from the earth or collecting in tar sands or tar pits. This seepage of crude oil from below a rock crust that originally pressurized it is a consequence of the relative light weight of oil. Because oil is lighter than rock, sand, and water, intrusion of such other materials can cause oil to migrate closer to the surface, through pockets and fractures in rock. The oil that is unable to rise to the surface may nonetheless form pockets, or reservoirs, below the surface at various depths.

Geologists have long since identified those conditions most favorable to the production of crude oil. Depressions that supported large lakes and seas, in regions of the world that were rich in the sunlight and temperature that support algae and zooplankton growth and eventual decay, created the environment for the prolific production of organic materials. If the rate of creation of dead organic material exceeded its ability to decompose, masses of organic materials could accumulate. Subsequent climatic and geological change that promoted the containment of masses of these decaying organic materials allow the decay to continue below the bottom of the sea, exposed to the heat of the earth's core and the pressure of mud, then rock and sand from above.

Tectonic shifts of land masses could move these reservoirs great distances over millions of years. As a consequence, oil is now found well inland, in areas that also contain the fossils of long-dried-up seas. For instance, ancient sea beds that are now the plains just east of the Rocky Mountains, from Alberta, Canada, through Wyoming, Utah, Colorado, Kansas, Oklahoma, and on to Texas, U.S.A., are all rich in oil, despite their great distances from today's oceans. These regions still show the fossils of shellfish, which demonstrates that they were once underwater and thousands of feet lower than their earlier locations.

The most significant pockets of oil are those that are "stored" below the surface of the earth or the bottom of a lake or ocean. Consequently, wells must be used to tap into the subsurface reservoirs to bring the oil to the surface.

Oil that is pressurized by the weight of water and rock above it, or by the spontaneous production of natural gas, may exhibit "natural lift" that easily brings the tapped oil to the surface. For instance, much of the oil first produced, from Texas and California, to Saudi Arabia, was naturally pressurized and was easy to tap and extract. As the reservoirs for this easy oil is depleted, the pressure decreases, and artificial lift, through pumps, become necessary.

The oil that first fueled the latter half of the industrial revolution relied on natural lift, followed by secondary lift. Geologists and oil companies balanced the cost of oil more difficult to extract with the cost of exploration of new fields with cheaper oil.

Oil was soon classified based on the cost of extraction and transportation to market. For instance, a reservoir that can be tapped and delivered to market for $10 per barrel is very profitable when the price of oil is $20 per barrel. More remote or oil more difficult to extract, might be barely economic when the prevailing price of oil is $20. Oil that is still more expensive to extract or bring to market would be deemed subeconomic until the price of crude oil rises accordingly.

Economic and subeconomic oil

As the world's most profitable "economic oil" is extracted first, and as scarcity pushes the price of crude oil up further, subeconomic oil becomes economically viable. The rising price of oil pushes wells deeper, farther flung, and more remote, and provides a greater incentive to extract more elusive oil in fields previously considered depleted.

An oil field can be abandoned if the reservoir becomes sufficiently depleted that pumping costs become prohibitive. Geologists have known for decades that this point of prohibitive cost is also related to the rate of extraction of oil. For instance, if oil permeates subsurface sand and rock above a reservoir, rapid extraction can deplete the reservoir before the rock and sand above can release its oil into the reservoir. The determination of a rate of flow that will allow oil to be fully released from the surrounding rock and sand can extend the production capacity of a reservoir, but only if extraction is sufficiently slow paced.

Good reservoir and extraction management, first through the primary processes of natural and artificial lift, can subsequently employ secondary processes that include the injection of water into the reservoir and surrounding rock to force the oil out. Further "tertiary" extraction can rely on the injection of steam and carbon dioxide, or, in the case of Arctic oil extraction, stranded natural gas, into the well to further pressurize and force the oil to the surface.

At one time, all oil extraction was primary. Now, in the United States, less than half the extraction is primary, while more sophisticated secondary and tertiary extraction represents the majority of oil production.

In addition, petroleum engineers have also learned to drill wells that no longer probe straight down through the ground or ocean floor. Horizontal and diagonal drilling can occur from one relatively small wellhead footprint, and can move miles downward and miles sideways. Specially engineered steerable drill bits also allow engineers to pilot wells through rock, avoid sand or problematic subsurface pockets, and tap into obscure reservoirs of oil that are detected through highly technical methods.

These methods that map the rock and sand below the surface or the earth or the bottom of the ocean can include the detection of materials through magnetic detectors or probed through seismic detectors. Engineers are offered a three-dimensional picture of the subsurface that indicate areas of high probability for hydrocarbons. Supercomputers processing these images can even optimize a drilling path and program a rig that can drill most efficiently to tap the reservoir.

As easy oil is extracted, and rising oil prices produce the incentives for oil exploration that is farther and deeper, colder and harsher, more engineering, technology, and investment is required to extract the oil.

Complex systems

As oil fields become more remote, and in harsher and more costly environments, the systems used to extract the oil become more sophisticated and hence more prone to failure.

The most obvious evidence of these shifts arising from scarcity and higher oil prices is the movement toward offshore oil over the past half century.

Offshore drilling actually dates back to crude oil drilling platforms and barges used to drill in shallow freshwater lakes in Ohio, and in the shallow salt waters off California in the United States in the late 19th century. By the 1930s, the Texas Oil Company (Texaco) developed barges to facilitate platform drilling in the Gulf of Mexico off the coast of Texas. By 1947, the precursor company to Anadarko Petroleum, the 25% owner with BP in the Macondo Prospect, was drilling offshore beyond sight of land.

Soon, technologies were developed that would allow oil exploration in water deeper than 100 feet. Fifty years ago, the Shell Oil Company pioneered an oil platform called Blue Water Rig No. 1 that was partially submerged. These semisubmersible rigs proved more stable and less prone to the disruptions of hurricanes and tropical storms that so frequently befall the Gulf of Mexico.

These drilling rigs and ships have become increasingly large, sophisticated, and expensive. The exploration and drilling operations are highly specialized, and are dominated by a number of exploration and drilling firms, including Kerr-McGee and Transocean Exploration. Oil companies such as BP, ExxonMobil, Chevron, and Shell Oil typically hire the rigs of these companies, at a rate upwards of $500,000 to $1,000,000 per day to discover, tap, and cap fields leased to them. These rigs are then moved to another part of the sea, or somewhere else in the world, as the subsea oil awaits production. Next, a production platform and crew will uncap the well and direct the oil into manifolds and pipes that form great subsea oil transportation networks.

These major offshore fields have been discovered and extensively tapped off in the Gulf of Mexico and the North Sea, off the coasts of Brazil, Nova Scotia, and Newfoundland, adjoining the West African countries of Nigeria and Angola, and, increasingly, in East Asia and Russia.

Some of the rigs that tap into these oils can stand almost 1,000 feet tall, can sit on the bottom of the ocean and rise to the surface, or may be moored and float above drilling networks that penetrate the earth at a point more than 10,000 feet below the surface of the ocean, as is the case of the Santos Basin off the coast of Brazil. And, the Tiber Well, in the Keathley Canyon Block of the Gulf of Mexico, contains reservoirs that are being drilled 35,000 feet deep.[11]

When drilling more than 10,000 feet below the ocean surface, where the pressure of water exceeds more than 300 atmospheres, or over 4,500 pounds per square inch, the operation is incredibly complex. These wells are far beyond the reach of direct human subsea diving, exploration, installation, and repair must be done by robots controlled by operators more than two miles away, or perhaps hundreds or thousands of miles away.

The challenges encountered in such drilling surpass even the challenges of space station repair. An astronaut repairing an installation outside the Space Station must work in a vacuum, or a pressure of zero atmospheres. Humans can work directly in environments from zero atmospheres to about 60 atmospheres using hard shell suits, or a maximum of 30 atmospheres using soft diving suits.

Only robotic Remotely Operated Vehicles, or R.O.V.s can work at depths three to six times the record depths humans have accomplished underwater. These robots come in direct contact with human technicians only when they are brought to the surface to be repaired or re-equipped. Otherwise, operators with an Internet connection located anywhere in the world can control a robot miles offshore and more than two miles deep. The technologies are some of the most exacting, sophisticated, and complicated of any technologies employed in any industry. The robots must work on drilling equipment that weighs hundreds of thousands of pounds and located at the highest pressures experienced in the natural earth, in complete darkness. If their operation leads to or cannot contain a failure, the potential for environmental damage is almost beyond compare.

To compound these challenges still further, companies are beginning to explore for oil underneath or in adjoining sea ice. Exploration efforts in the Barents Sea off Russia, and now in areas adjoining Elsmere Island off Canada's arctic region, must function in areas as inhospitable as are found anywhere on the planet.

Not unlike the conditions found at the much more accessible land-based oil fields in Prudhoe Bay, Alaska, these regions regularly encounter ambient winter temperatures often below −60 degrees Fahrenheit, or −51 degrees Celsius, with wind chill factors that bring the temperature much lower. However, while uncomfortable for the humans who must install the necessary equipment or repair the equipment in the case of an emergency, the robots used below the ocean or the sea ice operate at temperatures just above the freezing point of sea water.

One of the compounding challenges that occurs in such harsh conditions below the ocean's surface is that methane gas, often associated with oil, can form frozen hydrates at low temperatures and high pressures. These frozen hydrates, combined with the high pressure of the natural gas, and the natural pressures of some deep-sea wells create engineering challenges arising from freezing valves and burst pipes. These challenges exacerbate the consequences when absolutely everything does not go perfectly to plan.

Well-drilling basics

The industry has learned to accommodate many of these engineering challenges. Beyond the strength of materials that must be engineered to function in these harsh environments, their procedures are designed with multiple levels of redundancy.

All deep wells share a similar design. Each section of drilling shaft is screwed into the next to form a high strength union. As the hole is drilled, it is lined with a wider diameter casing that can act as a reinforced tube through which additional pipe can run. This outer casing is cemented in place so it bonds to the rock in which it is embedded. When the drilling is completed, the inner drilling shaft acts as the oil extraction pipe, with a "shoe track" valve placed at the bottom of the well to stop oil from migrating upward before the well is ready for production. Large solid rubber and steel gaskets seal the pipe in place in the center of the larger steel casing, and prevent sea water from penetrating the shaft at the seal on the bottom of the ocean floor.

In reservoirs pressurized by the weight of rock, sand, and water, or by associated natural gas, there is an added risk. This risk is shared on both land- and deep-sea-based wells. A self-pressurized well carries with it the risk of a blowout, should the pressure from below overcome the pipes and valves designed to contain it. As a last-chance, fail-safe mechanism, blowout preventers (BOPs) are employed. BOPs pinch the pipe closed with huge hydraulically operated rams, should the mechanical valves or the solid gaskets fail. One such ram is vulnerable to failure, making it prudent to employ two rams. However, should one ram be located just where one of the strong threaded sections of the pipe is located, it may be unable to pinch the pipe shut. This finite possibility calls for a three ram BOP to give both a high chance of pinching a pipe at a collapsible point with a sufficient degree of redundancy.

These multiple systems, of cement that contains the casing within the rock, a valve shoe that prevents flow from the well bottom, pipe strong enough to anticipate any foreseeable pressures, a gasket that can keep its integrity in such a harsh environment, a series of valves that can be shut as needed, and a BOP that can shut the whole well down if all else fails, are now employed in all deep-sea wells. Each of these elements fails on occasion. None of these multiple redundancies have failed simultaneously – until the BP Deepwater disaster.

Such challenges and expenses are undertaken, indeed demanded, solely because of our thirst for and the scarcity of oil. The use of engineering more sophisticated than might be used in space is profitable only if ever-growing, hydrocarbon-dependent economies are willing to pay a higher price for each barrel of oil extracted, and if these wells are large enough to sustain the immense exploration and extraction costs.

For instance, drilling at the Keathley Canyon block in the Gulf of Mexico, at a depth of almost seven miles, can cost upwards of a million dollars a day. These almost unfathomable expenses can result in huge rewards, though the reservoirs can contain upwards of 3 billion barrels, conservatively worth more than $200 billion. Such giant wells are expected to produce more than 400,000 barrels of oil equivalent a day, or daily revenue of almost $30 million, for decades. The $1 million daily drilling costs, and substantial risks, are small compared to the $30 million daily revenue the resulting well can produce.

Other risks

We have confined our analysis to the risk of exploration and extraction. There are additional, but unrelated, risks in the transportation of crude oil to refineries, through land-based, or ocean bottom pipe networks, and through oil tankers prone to oil spill accidents. During the refining process, the collection of flammable hydrocarbons of various volatilities, chemicals, and heat, and flames all create risks as hydrocarbons are separated into their various molecular weights and then recombined into mixtures that can burn in our engines and power plants. And, the transportation of some of the more highly volatile mixtures to our fuel stations, homes, factories, and power plants, is another risky process associated with the use of hydrocarbons. Finally, when these hydrocarbons are burned in our engines, homes, and factories, carbon dioxide is liberated to the atmosphere to once again begin a cycle measured in millions of years. Meanwhile, our burning of hydrocarbons inevitably produces carbon dioxide that blankets the earth and warms it through an insulating layer that passes lights from the sun to the earth's surface and reflects back some of the infrared light the earth would normally reradiate to space. The implications of the resulting global warming are difficult to determine. One cannot doubt the displacing effect that global warming will have on the economies of the world.

Nor can we afford to ignore the significant risks that we inevitably associate with a hydrocarbon-based economy.

Part II

The Uneasy Mix of Oil in Our Natural Environment

Oil spills have been occurring with a troubling frequency. If we are to disturb reservoirs of oil that have been held captive under rock for hundreds of millions of years, it is inevitable that accidents will happen. In this Part, we discuss these accidents, and the precarious technology that we use to draw oil to the surface and, in doing so, threaten to disperse oil in the environment. Safety in the oil industry in general and in oil exploration in particular are also discussed.

8
The Dirty Dozen before the Deepwater Horizon

The media described BP's Deepwater Horizon oil spill as the largest spill in history. From April 20, 2010, the day the site first began leaking oil uncontrollably, the media regularly reported each time the leak was estimated to have surpassed other significant spill milestones.

First, the spill was compared to the spill freshest in the minds of most U.S. observers; the Exxon Valdez catastrophe in the pristine Prince William Sound of Alaska more than two decades earlier. As the Deepwater Horizon site continued to spill oil into the Gulf of Mexico, it rose to the scale, and then surpassed, the Ixtoc I spill that occurred on the western side of the Gulf of Mexico before the Exxon Valdez spill. At that point, the media commonly reported that the Deepwater spill was the largest in history. However, a wartime spill created in the Persian Gulf in 1991 when Iraq invaded Kuwait remained larger.

With the historic perspective of the Persian Gulf spill duly noted, headlines began to qualify that the Deepwater Horizon spill in the Gulf of Mexico as the largest peacetime spill disaster. However, that assertion, too, proved false. In fact, the Deepwater Horizon spill is not even the largest spill in U.S. oil extraction history.

In this chapter, I document a short history of the worst and most catastrophic man-made spills over the past century of oil extraction.

The nature of man-made oil spills

Spills become notorious for one of at least three reasons. The spill may be:

1. large, in historic terms,
2. highly visible and easy for the world to observe, or
3. in a most sensitive ecological region.

There is one crucial feature all out-of-control spills share: a breach in a natural or man-made system causes a naturally pressurized reservoir of oil to spill spontaneously into its surrounding environment. The potential for, and the severity of, spills depends on the degree of pressurization of natural oil fields.

These fields of trapped oil can be pressurized by expanding gases, by the infiltration of materials or liquids into the reservoir, or by the sheer weight of rock, soil, or water from above. An untapped oil well is contained only by the integrity of rock, sand, or other materials that separate the reservoir from its surroundings. There can be natural breaches of such protecting layers, causing spontaneous releases of oil into the environment. The focus here, though, is the release of oil because of human interventions or failures.

The Lakeview Gusher – 9.4 million barrels released

The largest documented oil spill in history occurred on U.S. soil. The Lakeview Gusher in California in 1910 well exceeded the amount of oil that spewed into the Persian Gulf in the world's second largest spill, or in the third largest spill at the Deepwater Horizon rig.[12]

The Lakeview Gusher Number One is an excellent example of a man-made spill that created an out-of-control geyser of oil. Before it was contained, it had spewed 9.4 million barrels of oil. This release is almost twice as large as the next largest spill, the Gulf War spill into the Persian Gulf, and more than twice as large as the Deepwater Horizon spill into the Gulf of Mexico. And while the Gulf War spill took four months to contain, and the Deepwater Horizon spill took almost three months to shut down, the Lakeview Gusher rained oil upon an area north of San Francisco, California, for 18 months.

The Lakeview Gusher occurred at an oil reservoir known as the Midway-Sunset Oil Field in Taft, Kern County, California. At the time, the reservoir represented one of the largest oil reserves in the United States.

The Lakeview Oil Company was actually drilling for reserves of natural gas, a hydrocarbon that frequently is associated with oil deposits, and which often pressurizes oil fields. In its search for natural gas, Lakeview Oil was ill-prepared to contain significant amounts of oil, considered by the company to be a minor associated by-product of their natural gas exploration. In their previous exploration for natural gas, Lakeview Oil was not able to find any oil or gas, and had sold their interest to Union Oil Company of California (Unocal), since absorbed by and known as Union Oil. Unocal purchased the exploratory well not for the natural gas that was speculated to be below, but as a location for storage tanks in support of other wells in the vicinity. Because of the failure to find hydrocarbons and the need to use the property as a local storage site, the new owners issued word from their corporate office in Los Angeles to discontinue drilling.

However, some optimistic roughnecks disobeyed the home office order and continued to drill nonetheless. On March 14th or 15th of 1910, they tapped into the significant oil reservoir at a depth of 2,440 feet. The oil, highly pressurized in the reservoir by the natural gas deposits, streamed under high pressure up the drilling pipe and quickly overwhelmed the capacity of the ill-prepared drillers to contain the unexpected oil.[13]

While I will more fully discuss the engineering of oil drilling in later chapters, I note now that the pipe that guides the drilling shaft through previously drilled rock and soil also serves as the pathway for oil to come to the surface. This hollow steel pipe can range from a few inches in diameter with a quarter inch thick wall to upwards of twenty inches in diameter, or larger. The well pipe at the Lakeview well was of smaller size and inferior in design than those commonly employed today. The casing could not contain the pressure from the tapped crude oil, and lost its integrity, permitting the pressurized oil to blow out of the hole, much like a naturally occurring geyser.

Initially, the gushing oil was spewing oil 200 feet into the air and was reported to be releasing just under 19,000 barrels a day. At first, the gusher was spewing a similar amount of oil as was released early in the Deepwater Horizon spill. However, as the wellbore began to break down, the Lakeview Gusher was spewing considerably more oil than even the highest estimates for the BP discharge. As the pressure developed and the wellbore further deteriorated, the gusher released 90,000 barrels a day at its peak.

The Lakeview Gusher was literally creating a river of oil, contained only by the swift construction of sand bag barriers. Over its 18-month history, various containment attempts were made, from a large cap that was placed over the spill, but which subsequently blew out because of the high oil pressures, to a large sandbag–reinforced pond above the wellhead that created a sufficient column of oil to eventually contain the well. In its aftermath, a 60 acre lake of oil was formed.[14]

It was estimated that more than half of the 9.4 million barrels was recovered at the Lakeview Gusher site, while the remainder evaporated or was soaked into the soil.[15]

Local residents now celebrate the Lakeview Gusher with the annual "Gusher Days" events every March in Taft, California.

Gulf War Oil spill – 5 million barrels released

While the other spills documented here were produced quite by accident or negligence, the second largest man-made oil spill, and the largest ocean oil spill, was strategically created in an act of war.

In 1990, Iraq, under the leadership of Saddam Hussein, invaded Kuwait. The reason for the hostility was, not surprisingly, oil.

A portion of the border separating Iraq and Kuwait straddled various oil fields. While the etiquette and legality of sharing such binational oil fields is sometimes the subject of dispute, an indignant and increasingly aggressive Iraq accused Kuwait of cross-drilling into and pumping from the Iraqi side of the shared fields.

The binational dispute over oil simmered for months, with resolution becoming more and more elusive. Iraq responded by invading Kuwait and claiming the shared fields. Iraq had accurately calculated that the United States, long a partner in Kuwaiti oil production, would intervene. In an attempt to frustrate and complicate a U.S. invasion through the Sea Island oil terminal in Kuwait, the Iraq army opened valves that released oil from large oil tankers and storage tanks at the oil terminal harbor side.

The oil spill, which was initiated by the malicious opening of seaside valves on January 21, 1991, was partially disrupted by an aerial attack from American airplanes five days later. In the ensuing weeks and months of the conflict, other releases, from storage tanks, pipelines, and a damaged oil refinery, were successively contained.

Oil continued to be released into the Gulf until late May of 1991. Once the various sources of the spills were abated, it was initially estimated that 11 million barrels of oil were released. However, more refined subsequent estimates by the U.S. government lowered the estimates to between 4 and 6 million barrels.[16]

At its peak, the resulting oil slick in the Persian Gulf covered more than 4,000 square miles and was up to five inches thick in places. Various commentators claimed that the environment would suffer little from the damage, while others claimed significant multi-decade environmental damage. Half the oil had evaporated, another million barrels was recovered, and two to three million barrels swashed along the shore.[17]

Deepwater Horizon spill – 4.1 million barrels released

The various details of this spill will be described in subsequent chapters. The Flow Rate Technical Group, a scientific panel appointed by the federal government, estimates that 4.1 million barrels of oil was released into the Gulf of Mexico.

Ixtoc I – 3.45 million barrels released

The Gulf of Mexico is currently one of the world's most active oil exploration and extraction regions. Drilling is regulated by both the United States and Mexico for their respective offshore regions. It also represents a region of the world that contained two of the four largest oil spills in oil exploration and production history. Indeed, a 1,500 mile radius around Dallas, Texas, contained three of the world's four largest oil spills. Before the Deepwater Horizon spill, Ixtoc I represented the world's second largest well blowout in history, and the world's third largest spill. Ixtoc I also was uncannily similar to the circumstances and solutions that characterized the Deepwater Horizon spill more than three decades later.

The Government of Mexico had the responsibility to regulate the Ixtoc I reservoir, located in the Bay of Campeche, approximately 60 miles northwest of Ciudad Del Carmen of the Province of Campeche, in a relatively shallow depth of 150 feet.

The well was owned by the nationally owned drilling company Petróleos Mexicanos, called Pemex for short. In 1979, the company had been drilling over 2 miles below the ocean bottom and had been pumping drilling mud into the drill pipe to balance against reservoir-pressurized oil that would work its way to the surface. The use of drilling mud is an industry practice that helps to lubricate the drilling head, circulates drilling cuttings away from the drill point, and seals the gap between the drill shaft and casing, and between the casing and the previously drilled shaft.

The weight of the mud is essential for the safe drilling of oil from pressurized or gas-containing wells. The drilling mud must be sufficiently dense to provide a column of pressure to hold back oil and gas that would want to move up the shaft or up the gap between the shaft and the wellbore. Without mud of a proper density, an exploratory shaft that has yet been cemented in place can experience a blowout and release oil and natural gas to the surface.

The amount of drilling mud necessary to maintain well balance can vary, depending on the gap between the wellbore drilled and the steel casing inserted in place, any fissures or gaps that may exist in the rock being drilled, and the pressure of the oil below. The day before the catastrophic wellbore failure, the drilling tip hit a region of soft and fractured rock and sand. This sudden gap caused mud to flow into the fissures, thereby reducing the column pressure as mud was evacuated from the casing. The drilling team decided to remove the drill bit and reinsert the hollow drill column so that it could act as a carrier of mud directly to the soft strata it had encountered.

However, on removal of the drill column, the team was unable to contain the pressure from below. On June 3, 1979, the drilling platform Sedco 135-F ran out of drilling mud. The lack of sufficient mud caused oil and gas to flow up the wellbore column and precipitated a well blowout.

With a blowout, oil and natural gas will rise to the surface and may also rise up the drill shaft and gush out at the wellhead on the drilling platform. At first, nonvolatile drilling mud is spewed out. However, as the mud in the well casing is exhausted, a mixture of oil and gas gushes out of the drill shaft and onto the machinery surrounding the drill shaft. A spark, open flame, or sufficient heat from surrounding motors and pumps can ignite the flammable mixture of oil and gas and produce an intense fire that is constantly fed with pressurized fuel from below.

To prevent this dangerous condition, a blowout preventer (BOP) is typically inserted at the point in which the well casing penetrates the ocean floor. The BOP has one or more hydraulically operated rams that are activated to pinch the well casing shut. However, the drill shaft and casing is made up of a number of segments, each approximately 30 feet long. They are joined by a thread and collar that effectively more than doubles the thickness of the pipe. While these joints might represent six inches every thirty feet, one of these collars coincidentally aligned with the location of the pinching ram. As a consequence, the ram was unable to pinch the pipe shut.

The resulting gusher of mud, followed by oil and gas, on the Sedco 135-F platform resulted in an explosive condition. The natural gas and oil came into contact with machinery and caused a catastrophic fire on the platform. As the platform collapsed and sank to the bottom of the ocean, it took with it the associated drill shaft and lines. This collapse of the well structure severed the pipe containing oil and caused a catastrophic release of oil into the environment.[18]

At first, the spill was estimated to be releasing approximately 30,000 barrels per day. Attempts to pump mud into punctured oil lines reduced the flow by about a third in the following month. In August of 1979, two months after the accident, the pumping of "junk," a combination of steel and lead balls, into the well reduced flow by another third, to 10,000 barrels per day.

Meanwhile, Pemex drilled two shafts to intercept the reservoir and relieve the pressure. The first relief well intercepted the reservoir seven months after the accident. However, the well continued to release oil to the Gulf of Mexico for almost ten month, with final containment achieved on March 23,1980. By that point, an estimated 3.5 million barrels of oil were released into the Gulf.[19]

In response to the spill, Pemex tried to contain and recover some of the oil, and disperse the rest from airplanes flying over the slick. A dispersant is a chemical solution, made up primarily of hydrocarbons and propylene glycol, a nontoxic solution often used as an antifreeze, that combines with the released oil to produce much smaller globules able to more easily mix with the seawater.

This dispersal technique is only effective with relatively fresh oil that still contains a greater proportion of lighter hydrocarbon molecules. However, as these lighter molecules spontaneously evaporate, the dispersant proves ineffective on older and more weathered oil. Consequently, while Pemex commissioned almost 500 flights using the dispersant Corexit 9527, it did not use dispersants over U.S. territorial waters or, eventually, in Mexico waters at some distance from the wellhead.

The United States implemented a shoreline protection plan that would help protect the shoreline of the U.S. state of Texas, which adjoins Mexico. However, because Pemex is a nationally owned oil company, Mexico invoked sovereign immunity to shield its financial responsibility for cleanup of Texas waters and shoreline.

Scientific documentation of ecological damage to the nearby Mexico shoreline is incomplete. There was evidence of significant damage to larger sea animals such as turtles, fish, and octopi. There was better documentation of the effects in U.S. territorial waters. However, by the time the oil slicks penetrated U.S. waters more than 600 miles away, the effects were ameliorated.

The most significant seafood industry that was impacted depended upon the various species of shrimp harvested in the Gulf of Mexico. However, shrimp is thought to metabolize hydrocarbons relatively efficiently. Consequently, the U.S. Bureau of Land Management concluded:

> In spite of a massive intrusion of petroleum hydrocarbon pollutants from the Ixtoc I event into the study region of the South Texas Outer Continental Shelf during 1979–1980, no definitive damage can be associated with this or other known spillage events (e.g., Burmah Agate) on either the epibenthic commercial shrimp population (based on chemical evidence) or the benthic infaunal community. Such conclusions have no bearing on intertidal or littoral communities, which were not the subject of this study.[20]

The incomplete and speculative record of damages from this spill makes apparent one need. A routine part of drilling in any sensitive area should be a baseline environmental assessment and inventory. In the event of a subsequent spill, this inventory can then be used to better assess long-term effects of a catastrophic spill.

The Atlantic Empress – 2.1 million barrels released

In 1979, the Atlantic Empress, an oil tanker of the class Very Large Crude Carrier (VLCC), was involved in a collision with another VLCC, the Aegean Captain, off the coast of the southern Caribbean island-state of Trinidad and Tobago. It represents the largest ship-based oil spill to date.

The Atlantic Empress tore the bow off of the Aegean Captain, and resulted in the loss of 26 lives. The Aegean Captain successfully controlled the resulting fire on it ship, and ultimately spilled only modest amounts of oil as it was towed to Curacao. The oil that was spilled was sprayed with dispersants by the tugboats that returned the Aegean Captain to port.

The collision with Atlantic Empress suffered more damage, though. It was initially towed toward open sea for fear that a significant spill would damage nearby islands. Fire boats sprayed it with water, as the ship was pulled by tugboats to open waters, with a burning slick of oil in its trail. Four or five days after the July 19 collision and fire, a series of explosions rocked the ship, with an even larger explosion erupting on July 29. By August 2, the ship was beginning to list badly, and the tugboat unhooked. By August 3 of 1979, only a burning slick remained.

The ship ultimately lost 2.1 million barrels of oil. While no impact study was performed, no significant shoreline pollution was noted. Instead, media attention soon diverted to the emerging Ixtoc I spill in the Gulf of Mexico.[21]

While the Atlantic Empress spill is the largest ship-based oil spill to date, it does not represent the largest potential ship-based oil spill. The MV Atlantic Empress was a VLCC class oil tanker, classified with a capacity of 250,000 deadweight tons (DWT, or the equivalent of 2 million barrels of oil. The largest tankers now constitute the ULCC (ultra large Crude Carrier) class, with a capacity of 320,000 DWT or greater. The largest of these ULCC supertankers, the Seawise Giant, was built in 1979, and operated as a tanker until 2004.

The four largest supertankers that continue to ply the seas are sister ships. The TI Asia, TI Europe, TI Oceania, and TI Africa were built in 2002 and 2003 with double hulls for additional safety, following recommendations that were promulgated in the aftermath of the Exxon Valdez spill, to be described later.

These ships are so large that they cannot navigate the English Channel. They are able to transport nearly 3.2 million barrels and are about the same length as the Empire State Building. The catastrophic loss of one of these ships laden with oil would constitute the fifth largest man-made spill in history. Its size would be just over a third the size of the Lakeview Gusher in the United States, about three fourths the size of the Gulf War spill and the Deepwater Horizon spill, and about the same size of the Ixtoc I spill. Such a catastrophe would be about 50% larger than that spilled by the collision of the Atlantic Empress and the Aegean Captain VLCC supertankers. However, the offshore collision of two such ULCC tankers would easily constitute the largest oil spill in a century and the second largest spill in history.

Such large-scale sea navigation of oil carriers can be placed in some perspective. In 2007, the United States used 20.68 million barrels per day, or the capacity of the equivalent of almost 7 of the world's largest supertankers each day. China, with a consumption of 7.6 million barrels, would require the capacity of more than 2 ULCC class supertankers per day. Japan would require more than 10 such tankers per week, while Russia, India, and Germany each require the load of almost one such supertanker each day.[22]

Fergana Valley – 2.1 million barrels released

The Fergana Valley bordering Uzbekistan and Kyrgyzstan has an extremely dense array of wells, farms, and people. It has been intensively farmed and inhabited since prehistory. Rich deposits of oil were discovered in this region, and competed with agricultural land under a poorly regulated industrial infrastructure. These factors conspired to help create a major land-based spill.

The Fergana Valley spill, also known as the Mingbulak oil spill, was a large land-based spill at the Mingbulak oil field of the Fergana Valley in Uzbekistan. Equipment failure at a production well precipitated a blowout that spewed 88 million gallons of crude oil into the valley. It constituted the worst oil spill in the history of Asia, at more than 2 million barrels. The resulting fire from the blowout burned for two months, and consumed a good share of the oil.

The estimate of 2.1 million barrels is based on the amount of oil contained by temporary dykes at the wellhead. The blowout eventually depleted the well, and allowed the oil release to abate naturally.

ABT Summer – up to 1.9 million barrels released

The ABT Summer, a VLCC class supertanker had rounded the southern tip of Africa laden with 1.9 million barrels of heavy crude oil from Iran, en route to Rotterdam, the Netherlands. About 900 miles off the coast of Angola, an onboard explosion on May 28, 1991 killed five crew members and set the ship ablaze. The remaining 27 crew members were evacuated before the ship sank 4 days later.[23]

It is difficult to estimate how much oil was released to the sea and how much was consumed by the intense supertanker blaze. However, the entire cargo of more than 1.9 million barrels of oil was lost. While an oil slick of 80 square miles was created, the open seas broke up the slick rather swiftly, and likely with relatively little measured long-term environmental impact.

Nowruz Field Platform – 1.9 million barrels released

Certainly, if one could rank oil spill locations based on a history of repeated accidents or calamities, the Nowruz Field Platform in Iran stands above the rest.

The Nowruz Oil Field is one of the premier Persian Gulf fields. It also lies in close proximity to a major shipping channel and in the territory of Iran, a nation bound up in a number of regional conflicts.

On February 10, 1983 an oil tanker servicing the field collided with one of the well platforms. The collision forced evacuation of the platform as it began to list. The combination of the severe list, wave action, and corrosion caused the riser pipe connecting the wellhead at the bottom of the sea and the platform to collapse. This riser collapse, not unlike that experienced by the Ixtoc I and the Deepwater Horizon spills, resulted in a discharge rate estimated to be 1,500 barrels per day.

Conflict between Iran and Iraq prevented any capping effort in this war zone. A subsequent air attack by Iraqi jets in March of 1983 further hampered capping efforts of a, by then, blazing and growing oil slick. It took six months following the air attack, and the loss of eleven lives, for the well to be capped.

In the meantime, another platform in the field was also attacked by Iraqi helicopters in March. This second attack induced an additional spill at a discharge rate of 5,000 barrels per day. This second discharge began to deplete, but was still producing 1,500 barrels of oil per day for two additional years before the well was successfully capped. By the time this well was capped in May of 1985, nine more men lost their lives, and a combined 1.9 million barrels had spilled into the Persian Gulf.[24]

Castillo de Bellver – 1.85 million barrels released

The Castillo de Bellver was a VLCC class supertanker that carried 1.85 million barrels of light crude oil as it rounded Cape Town, South Africa on August 6, 1983. The ship caught fire, lost power and control, and went adrift toward the shore of South Africa. It subsequently broke into two pieces. The stern section capsized with about half of the ship's oil in its holding tanks, 24 miles off the coast. Meanwhile, the bow section was towed away from the coast and was sunk with controlled explosive charges.

The oil was eventually released in these two locations, veering first toward shore but subsequently taken out to sea by the wind and current. While 1,500 gannets were oiled at the height of their breeding season, and a number of seals were present in the vicinity of the oil dispersant activity, little mitigation was performed, and environmental damages were difficult to document as the slick was brought out to the open sea.[25]

The Castillo de Bellver remains the largest spill in South African history, and occurred in a regionally significant fishery. Like many sunken tanker incidences, the long-term effects remain questionable as all the oil holding tanks may not have emptied as the ship sank. The combination of corrosion and pressure could subsequently allow oil to seep or to be released more catastrophically at some time in the future.

Amoco Cadiz – 1.6 million barrels released

Earlier in this chapter, it was noted that spills become notorious for a number of reasons. First, the size of the spill can be large, compared to other man-made spills in the last century. Second, the spill may be in an area for all to see, as compared to a spill in the open sea and far from the scrutiny of cameras and reporters. Or, third, the spill may be in a most sensitive environment that commands our collective attention. The Amoco Cadiz oil tanker spill, off the coast of France, met the last two of these criteria.

With a release of 1.6 million barrels of oil, the Amoco Cadiz was considered a very large crude carrier (VLCC) supertanker in 1978. On March 16, the Liberian registered, Spanish built supertanker owned by Amoco was carrying Persian Gulf oil enroute to Rotterdam, the Netherlands, with a stop at Lyme Bay, Great Britain.

As the ship passed Brittany and entered the English Channel, gale force conditions caused the ship to be battered by waves. One particularly heavy wave struck the ship's steering rudder and damaged the studs that controlled the steering gear attached to the rudder. The damaged studs induced a leak of hydraulic fluid that soon prevented the ship's helm from having any steering control.

Half an hour after the damage, the ship's captain ordered a message to be sent to other vessels in the area. The radio message "no longer maneuverable" was not accompanied with any specific request for tugboat assistance for another inexplicable hour.

Shortly after the Amoco Cadiz called for tugboat assistance, the tug Pacific responded under the provisions of maritime law called "Lloyd's Open Form." Such LOF assistance is offered in exchange for some salvage rights as mediated by the Lloyd's of London insurance organization. Since the ship and its cargo was valued at approximately $40 million, the captain of a successful salvage could be rewarded a large amount. However, the captain of the Amoco Cadiz refused the terms. The two respective ship captains continued to haggle over terms for hours, even as their crews began to cooperate to rescue the ship at peril.

While the rescue tug arrived at 12:20 p.m., a tug line could not be attached successfully until 2:00 p.m. The rough seas made attachment difficult, and caused the line to subsequently break. It was not until just after 9:00 p.m. that the line was successfully and robustly attached. By then it was too late. The Amoco Cadiz ran aground off the coast of France moments later for the first time. The engines were flooded, the ship continued to be battered by Gale Force 10 winds, and the ship ran solidly aground half an hour later.

The crew was rescued by French navy helicopters, while the captain and one crew member insisted on staying aboard for another seven hours. Five hours after the captain and the remaining crew member abandoned ship, it broke up, and released its cargo of 1.6 million barrels of light crude oil from Saudi Arabia and Iran. The gale force winds and high seas prevented rescuers from pumping any oil from the ship before it broke up.

The resulting slick was 12 miles long and soiled 200 miles of beach, at times to a depth of 20 inches. The significance of the spill lies with the population and history of the Brittany region and the most significant documented loss of marine life of any spill to that date. Millions of shallow bottom-dwelling sea animals, 20,000 birds, and almost 20,000,000 pounds of oysters were lost. Crustaceans, and echinoderms such as sea urchins almost completely disappeared, although other species returned within a year. As we subsequently discovered following the Exxon Valdez disaster a little more than a decade later, the aggressive oil scrubbing of shore rock with pressure and steam caused even further habitat degradation.

The spill's visibility, in environmentally sensitive waters, allowed some of the most complete studies of the environmental effects of oil spills to date. In addition, the French government calculated the economic damages to total $250 million in direct consequences to the fisheries and tourism. They presented a claim of US$2 billion to Amoco in U.S. courts, and were subsequently awarded six cents on the dollar, or $120 million, 12 years later.

Amoco would be absorbed by BP later in the decade in which the court proceedings were finally settled.

MT Haven – 1.06 million barrels released

The last spill to exceed a million barrels was the wreck of the MT Haven, the sister ship to the Amoco Cadiz. The Cyprus registered, Amoco owned ship, built just before the Amoco Cadiz in Cadiz, Spain, was a VLCC supertanker that had been leased to the Greek shipping company Troodos. While unloading crude oil at a floating platform just off the coast of Genoa, Italy in 1991, there was an explosion that immediately killed five crewmen. A sixth man died later.

As Italian emergency responders tried to quench the fireball, explosions continued to ring out. The authorities managed to pull the ship from the platform to a point where it impinged on the sea bottom off the coast. More than half of the oil was successfully pumped from intact holding tanks as the ship lay on the bottom of the sea. The bulk of the ship's oil was also burned off in the fire and explosions, and the balance, an estimated 250,000 barrels of oil slowly released over the next decade and polluted the nearby Italian and French coasts. The shipwreck remains a popular local attraction for divers.

The Greek father and son who had leased the ship were prosecuted but later acquitted for manslaughter in the deaths of six men for failure to maintain the ship adequately. Italian officials and representatives of the merchant officers' union were embittered by their acquittals and the failure of the courts to award civil damages.

Torrey Canyon spill – 675 million barrels released

The Torrey Canyon spill, while not as large as the other more recent spills involving larger supertankers or more risky well drilling scenarios, is perhaps the first spill of the modern era of media and environmental awareness. The ship was owned by Barracuda Tanker Corporation, but was leased to British Petroleum, the company since renamed BP. The Liberian registered ship was originally built as a 60,000 ton tanker in the United States in 1959, but was subsequently doubled in size in Japan. While half the size of the VLCC supertanker class, it was nonetheless the largest shipwreck in history when it went down off the coast of Cornwall, England in March of 1967. The Torrey Canyon spill was the first major oil spill in the supertanker era.

The ship had left filled with Gulf oil from Kuwait and was destined for Milford Haven, England, with a short stop in the Canary Islands. The captain, in his haste, took a shortcut to save time on his way into Milford Haven so that he could make the high tide. As a consequence of this diversion, the ship was left impaled on the Seven Stones reef.

In the prevailing maritime navigation environment at that time, ships had to deliver cargo anywhere in the world but without the sophistication of satellite-based mapping and navigation, a long distance ship had to carry a great number of charts. To save space, the captain would choose charts that could individually navigate a larger area, albeit using a smaller scale and reduced map features. Consequently, the maps used by the captain did not reveal sufficient detail to make safe his unfortunate detour. In addition, an inexperienced helmsman did not realize that the ship's wheel had inadvertently been left between autopilot and manual pilot, and the ship had been sailing with no control.

When the captain realized the dangerous condition, it was too late. The ship soon ran aground. Despite repeated efforts by a Dutch salvage team to tow the ship off the reef, the ship remained impaled on the rocks.

The ship soon began to leak oil from ruptures in its hull. Authorities tried to disperse the oil with detergent, but the slick continued to worsen. The British cabinet under Prime Minister Harold Wilson ordered the ship to be set ablaze in an effort to burn off the remaining oil. Meanwhile, booms surrounding the ship were ineffective because of the high seas.

In its aftermath, the oil release from the Torrey Canyon spoiled almost 200 miles of French and English coastline on both sides of the English Channel. Fish within a 75 mile radius, and 15,000 seabirds that fell within the almost 300 square mile slick were lost. Some of the damage occurred because of the clumsy and ill-advised use of detergents in this first major oil spill. Much of what we know now in our handling of spills arose because of the failures in the Torrey Canyon spill.

The difficulty encountered as the British and French government sued for damages also forced a change in the law governing oil spills in international waters. Prior to the spill, plaintiffs had to demonstrate, by a preponderance of the evidence, that the ship owners and its agents behaved negligently. The subsequent Civil Liability Convention of 1969 imposed strict liability on ship owners. Subsequently, plaintiffs no longer had to prove negligence, with liability automatically assumed by ship owners. The spill also motivated the International Convention for the Prevention of Pollution from Ships in 1973.

The Greenpoint spill – 55 million barrels released

All of the spills described occurred in a dramatic fashion. This final spill was smaller in size, but much more insidious in its invisible spread. While more than twice the size of the Exxon Valdez spill, an eighth the size of the Deepwater Horizon spill, and one-eighteenth the size of the Lakeview Gusher, it represents the third largest spill in U.S. history, the longest running spill in history, and occurred in one of the world's most densely populated regions. The Greenpoint spill, in the Greenpoint neighborhood of Brooklyn, New York, released between .4 and .7 million barrels of oil over decades of soil and groundwater pollution.

The Greenpoint area hosted dozens of oil processing plants dating back to the 1840s. Many of these early refineries became part of the notorious Standard Oil (S.O.) near-monopoly that eventually morphed into Esso and then Exxon. By 1993, the area had been employed as an oil distribution terminal for Amoco (later absorbed by BP), and Paragon, later absorbed by Chevron.

While it was determined that the area had been spilling oil intensively for much of the 1900s, the extent of the leak was not discovered until 1978. A Coast Guard patrol detected a plume of oil flowing into a neighborhood creek. A subsequent investigation showed that the spill of upwards of 700,000 barrels had contaminated at least 100 acres of soil.

Pumps operated by ExxonMobil, BP, and ChevronTexaco continue to remove oil from the site. The State of New York Department of Environmental Conservation claimed in 2006 that approximately 200,000 barrels, or almost the amount of oil spilled by the Exxon Valdez, have been recovered. A report issued by the U.S. Environmental Protection Agency determined the 700,000 barrel contamination figure, which places the spill of a size three times larger than the Exxon Valdez.

In 2005, residents near the contamination site filed a lawsuit against ExxonMobil, BP, and Chevron, claiming health problems arising from elevated levels of the more volatile hydrocarbon molecules. The primary defendant, ExxonMobil claims that Paragon Oil, a company long ago absorbed by Chevron, was named the responsible party.[26] Exxon continues to deny liability for the claims.

Less-than-Honorable Mention – The Santa Barbara blowout – 60,000 – 100,000 barrels

California had a long history of on-land oil production, and still holds to this day the record for the largest oil spill. However, as these sites were discovered and exploited, there was increasing pressure to drill off of California's shore. As early as the1890s, rigs were constructed on the beaches of Santa Barbara Channel, and were competing with the noncompatible growth of local tourism and health spas. However, the value of oil was compelling, and approvals were granted for offshore exploration.

The first mammoth oil spill to occur in California, and the second largest offshore oil spill in U.S. history, next to the BP Macondo spill, occurred off of Santa Barbara on Platform A, in January of 1969. Union Oil-owned Platform A, one of a dozen platforms off the shore of California, had already drilled four wells, and had reached its planned depth of almost 3,500 feet on a fifth well. Most of the well remained uncased with a thick steel sheathe. As the crew was pulling out the drill bit, but before a blowout preventer had been fully installed to seal the well, oil and natural gas began gushing from the drill pipe and into the air above the platform.

As a last resort, the crew plunged the drill bit back into the hole and pinced off the pipe with blind rams, not unlike those used on modern blowout preventers. Once this hole was contained, workers noticed oil and gas bubbling to the surface of the ocean at a distance away from the platform. High pressure oil and gas was making its way through the ocean floor. Because the platform crew had not fully inserted casing to a depth required by federal regulations, the oil and gas below had migrated up the wellbore and began to seep into cracks and fissures in the porous sandstone within the first couple of hundred feet of the ocean bottom. This seepage continued for a year and a half.

In the aftermath of the spill, 3,686 seabirds died from an 800 square mile oil slick that oiled 35 miles of coastline.[27] Following the blowout, the California State Lands Commission banned new drilling within three miles of the California shore. In 1981, the U.S. Congress followed this with a ban on offshore drilling in continental shelf waters, except in the Gulf of Mexico and parts of Alaska. Congress let the ban expire in 2008.

I append in a separate chapter one spill that adds to this list of the dirty dozen. The Exxon Valdez is not nearly as large as the others, almost all of which exceed a million barrels of oil discharged to land or sea. But, while the Exxon Valdez spilled only 250,000 barrels, various environmental, legal, and media factors raise it to a level of much greater scrutiny and interest.

9
The Case of the Exxon Valdez

Lastly, I explore a spill that was less than half the size of the smallest spill documented previously. However, despite its smaller size, it was, until the Deepwater Horizon tragedy, the most significant spill in the minds of the American public, and has created the backdrop and the tone for media coverage ever since.

The wreck of the Exxon Valdez supertanker off the coast of the State of Alaska, United States, resembles the Deepwater Horizon incident in two important ways. It occurred in an area of significant environmental sensitivity and it was highly visible to an American public that was just becoming acquainted with the modern 24-hour news cycle. It also defined much of the legal and regulatory network that was subsequently applied to BP, the responsible party in the Deepwater Horizon spill.

A Panamanian registered VLCC supertanker the Exxon Valdez was owned and operated by the Exxon oil company. Oil was pumped from wells in the huge Prudhoe Bay oil fields on the North Slope of Alaska on the Arctic Ocean, and travelled to the Port of Valdez through the Alyeska pipeline, a 36 inch diameter pipe that travels 800 miles through some of the most rugged and earthquake-prone terrain on earth. Between 1.5 million and 3.5 million barrels per day flow through that pipeline and fill up to a supertanker each day at Valdez, the pipeline's southern terminus. From there, supertankers regularly shuttle oil another 1,200 miles to refineries on the west coast of the contiguous United States.

Alaska is a land of superlatives and Alaskan oil is no different. The industry epitomizes the challenge of modern hydrocarbon production and consumption. Prudhoe Bay is North America's largest oil field. At a latitude of 70.3 degrees, the field lies well within the Arctic Circle and less than 1,200 nautical miles from the North Pole. It is in an area that has three months of total darkness every winter, and four months of nothing but light every summer.

Despite continuous production and an aggregate extraction of 11 billion barrels at the field since the completion of the Alyeska Pipeline in 1977, BP, the principal well operator and co-owner with ConocoPhillips and ExxonMobil, estimated in 2006 that 2 billion barrels of oil still remain.[28] At an oil price expected to hover around US$100 per barrel for the coming years, the oil remaining in this field represents a value of approximately $200 billion.

The Prudhoe Bay oil field experiences some of the harshest climate conditions in the world. With winter temperatures regularly falling below −40 degrees Fahrenheit and Celsius,[29] and wind chill factors bringing the equivalent temperature for human activity as low as −100 degrees Fahrenheit, exploration and extraction is only for the hardiest. However, while humans have learned to adapt and work in such harsh conditions, few mechanical processes are designed to reliably function in such cold weather. Humans find themselves modifying and repairing machines to operate in an environment that they would otherwise avoid. They do so, of course, because the high and increasing demand for oil makes it worth their while.

Valdez was chosen as the southern terminus for the Alyeska Pipeline because its harbor and the adjoining coast of Alaska and British Columbia, Canada, are ice-free for much of the year. The navigation route is also reasonably protected and not prone to severe seas. However, there are some well-understood navigational hazards that must be avoided, most notoriously the Bligh Reef in Prince William Sound at the exit of Valdez Harbor.

On March 23, 1989, Shipmaster Joseph Hazelwood departed the Valdez terminal with 1.3 million barrels of oil aboard the Exxon Valdez. March is just beyond the peak of ice problems in Prince William Sound, so it was not unusual to have to navigate around some relatively small icebergs at that time of year. On this evening, the usual outbound shipping lane contained icebergs. As was routine, Hazelwood secured permission by radio with the Coast Guard to deviate from the shipping lane. Once the deviation was initiated, Hazelwood retired to his quarters after activating the ships autopilot and leaving control of the wheelhouse to his Third Mate, Gregory Cousins. An hour later, just past midnight on March 24, the ship struck Bligh Reef.

It was commonly reported that the fault lay with a negligent captain who had been drinking the evening before, and had been sleeping off the alcohol at the time of the accident. However, complex systems with multiple redundancies rarely fail for a single reason, as we shall see with our investigation of the Deepwater Horizon spill.

The Exxon Valdez had been launched by the National Steel and Shipbuilding Company in San Diego, California barely two years earlier. However, it had been equipped with a Raycas sonar depth measurement system that had malfunctioned a year before the accident and had not been repaired because of the cost. When the tanker confined itself to established shipping lanes, as is typical, this failure would be of no consequence. However, the deviation to avoid icebergs in the outbound shipping channel made the ship vulnerable to the failed navigation equipment and more reliant on navigational charts.

There had been recommendations that the shipping channels around Prince William Sound should have been continuously monitored with a state-of-the-art ice detection system. Had this advice been heeded, the ship's captain may have been able to make a better-educated judgment regarding the necessity of a departure from the shipping channel. Nonetheless, crews were also under the false assumption that the Coast Guard actively tracked ships around Bligh Reef, much like air traffic controllers monitor air traffic to ensure safety and a redundancy system for collision avoidance.[30]

As another cost saving measure, Exxon had cut by half the size of its ships' crews over the previous decade. The staff reduction necessitated 12-hour shifts, and made the crew more dependent on technology. Further compounding fatigue, the crew on that day had not received their legally mandated rest before their shift began.

Coast Guard inspections and spill response plans in the Port of Valdez had also become complacent, perhaps as a consequence of overconfidence from a long period of accident-free shipping. The Exxon Valdez inspection cycle had not detected and mandated a repair of the critical sonar depth measuring system.

Finally, Joseph Hazelwood was known by Exxon to have a drinking problem. At the time of the accident, his driver's license had been suspended by the State of New York as a consequence of driving under the influence of alcohol in September of 1988. Previous infractions in 1984 and 1985 had induced him to enter a dependency rehabilitation program in 1985. Exxon had even provided him 90 days of leave to attend Alcoholics Anonymous.[31] A lax and cost conscious corporate culture had avoided dealing more effectively with this potentially dangerous factor.

In the ensuing criminal trial of Captain Hazelwood, an Alaskan court cleared him of a charge of operating a vessel while intoxicated.

However, he was convicted of the misdemeanor charge of negligent discharge of oil, sentenced to 1,000 hours of community service, and fined $50,000.

The spill

It is unlikely that any one of these factors solely led to the reef impalement and spill of the Exxon Valdez. Rather, like most major releases of oil, a series of unfortunate circumstances conspired to create the largest release of oil in U.S. waters.

Once grounded on Bligh Reef, the captain returned to the helm and tried repeatedly to free the ship from the reef. In the process, the reef gashed the single-skinned hull and punctured a portion of the oil holds. A State of Alaska investigation reported that the ship released 250,000 barrels, or a little less than a fifth of its Alaskan oil cargo, although other groups challenge that the true number may be two to three times higher than this amount.[32]

The Alyeska Pipeline Company had created a spill response barge as a contingency for such a spill. However, the barge equipment had not been maintained or restocked. Therefore, the initial response to the spill proved inadequate. It relied primarily on dispersants to reduce the size of oil globules so that they would mix easier with water, surfactants that would prevent the oil from remaining on the surface of the seawater, and solvents that would thin the oil for easier mixing with water. Burning of oil on the surface was also used initially.

Subsequent cleanup relied on what is called mechanical methods. As the slicks came to shore, high-pressure water streams and steam was used to clean the rocky shore. The oil washed off the rocks would then be collected by booms strung along the shoreline. However, this mechanical removal technique also forced the oil further down under the rocks, where some of it remains today. The mechanical method also cleansed the rocks of the organisms that would have helped degrade the oil naturally, and that would allow repopulation of the natural habitat after a more successful cleanup.

Despite these efforts, which were widely regarded as slow and ineffective, there had been little preparation or advance research on large-scale oil remediation in that harsh climate and rocky shoreline. A recent report claimed that only 10% of the oil was recovered. An estimated 1,500 miles of shoreline was oiled, many hundreds of thousands of marine animals and shorebirds died, and plankton and smaller fish lower down on the food chain have yet to return in previous numbers, thereby decreasing numbers of larger fish, as well.[33]

The visual picture of thousands of dead birds and sea mammals, ranging from otters to orcas, against a backdrop of otherwise pristine natural beauty, attracted global media attention. Given the low average air and water temperatures and the relative protection of some of the affected coastline, the environment has yet to fully recover, more than two decades later. Scientists at the University of North Carolina estimate that the habitat may take yet another decade to return to its previous state.[34]

In addition to numerous ecological studies, other economic studies have attempted to measure the effects on recreation, fisheries, tourism, and the loss of "existence value" that represents a society's collective valuation of a pristine environment.[35]

The spill became the most studied environmental disaster to date, and put Exxon on the defense at the time

The fallout

The State of Alaska has responded to the Exxon Valdez disaster by requiring two tugboats to escort loaded tankers past the reef. Now, one of the escort vehicles is also able to serve as a spill emergency response vehicle.

Following the refloating and repair of the Exxon Valdez, laws were passed that required double-hulled tankers in U.S. waters. The Exxon Valdez became obsolete as a supertanker in U.S. waters. Despite the fact that it had been built only three years before running aground in Alaska, it was subsequently renamed the S/R Mediterranean after $30 million of repairs. Currently, Panamanian registered and owned by a Hong Kong company, it is now an ore carrier renamed the Dong Fang Ocean.

The Exxon Valdez incident has also helped define the modern corporate culture in the eyes of the public. Perhaps like no other corporate malfeasance but for that other Texas corporation, Enron, Exxon's response in the matter of the Exxon Valdez has created the lens by which corporate environmental or social irresponsibility has been judged ever since.

Exxon learned a number of things from the Exxon Valdez experience, though. It conducted a strategy out of the public eye that would first allow it to successfully negotiate with the government to pay civil rather than criminal penalties for its transgression. By doing so, these penalties could be written off as business expenses in its taxes.

Exxon realized, too, that, by stretching litigation with thousands of Alaskan plaintiffs over a couple of decades, it could whittle down its liability from billions to mere millions. In addition, it discovered that few pay close attention once the spectacle is over and the media has moved off to another story. Finally, it realized that it must embrace a much higher standard for safety because the public will be much more punishing the second time around. As a consequence, Exxon strives to demonstrate a strong commitment to corporate safety ever since.

Negotiated liability and the cost of cleanup

It was inevitable that, while the response team from Alyeska and Exxon was coordinating with the State of Alaska in the cleanup of the oil spill, the legal team was busy developing a legal strategy that would minimize the ultimate spill cost to shareholders.

The actual cleanup costs were typically small compared to the various economic damages associated with a spill in a commercially active area. Exxon would ultimately be responsible for cleanup cost that totaled in excess of $2 billion, according to their reports. This estimate is based on its own cleanup efforts, those performed by the State of Alaska and reimbursed by the State, and a fund created by Exxon to fund the cleanup and environmental rehabilitation by third parties.

Economic damages

In subsequent litigation, any court-imposed punitive damages or fines are usually proportional to the economic damages suffered by plaintiffs or the state. The punishment is determined so that it might deter others from acting in an irresponsible manner in a similar future circumstance. Optimal deterrence must be designed to ensure that corporations mitigate in advance the costs its decisions impose on other parties. Responsible parties need little deterrence beyond the financial responsibility of their actions. Irresponsible parties must be further deterred in proportion both in the damages they incur on others and on their pattern of negligence.

The most negligent corporations can be deemed so criminally negligent that they violate the sensibilities of society. The stigma of criminal sanctions is severe, as must be the standard of proof and guilt necessary to prosecute such transgressions. Such criminal fines typically represent a multiple of the actual economic damages incurred. On the other hand, civil punitive sanctions, over and above court-imposed economic damages, are tax deductible. Civil negligence is considered only a violation of the economic rights of plaintiffs, not society as a whole. These costs are then considered a cost of doing business.

With economic damages, to fishermen, tourism-related businesses, and property owners in Prince William Sound, expected to rise to hundreds of millions of dollars, Exxon lawyers correctly surmised that civil sanctions could amount to billions of dollars, and criminal sanctions, if imposed, could be larger still. Exxon very swiftly negotiated with the Federal Government to secure an agreement that subsequent proceedings would be conducted in civil rather than criminal court. By doing so, they likely saved Exxon hundreds of millions of dollars in criminal fines and avoided a tax liability of additional hundreds of millions of dollars. These fines could not be considered by the corporate tax accountants as a regular and tax-deductible cost of doing business. Hence, if a corporation must pay 38% of their profits in federal taxes, it cannot avoid 38% of the costs of criminal sanctions on its tax return.

Exxon was not entirely successful in avoiding criminal sanctions. As a consequence of the 260,000 barrel spill, it paid $125 million in criminal penalties and $900 million in civil penalties. Because of the reduced tax liability arising from a write-off of the civil penalties, U.S. taxpayers ultimately paid a little less than $400 million of these penalties.

While state and federal entities took over the cleanup efforts early on, the courts were used in an attempt to recover damages from Exxon. The first major suit, Baker v. Exxon, in front of a jury in Anchorage, Alaska, awarded the plaintiffs damages of $287 million, and punitive damages of $5 billion, calculated to equal one year's profits for Exxon. Interestingly, Exxon's hedge against the claim also created the first "credit default swap," an instrument that a decade later became ubiquitous, wreaked havoc on world financial markets and contributed to the most significant downfall of world economies since the Great Depression.

Exxon appealed the decision in Alaska court that imposed $5 billion in punitive damages. In its appeal, Exxon changed the venue from Alaska to Seattle, Washington, as it appealed to the 9th Circuit Court of Appeals, and finally to Washington, D. C. and to the U.S. Supreme Court. Almost 20 years after the spill, Exxon prevailed. On June 25, 2008, on behalf of the court majority, Justice David Souter concluded that Exxon was "worse than negligent but less than malicious" in its conduct preceding the spill.

ExxonMobil was eventually ordered to pay $507.5 million to Alaska Natives, business and property owners, and fishermen damaged by the spill. With an interest of 5.9% imposed on the amount owed, tolled since 1996, the punitive damages award amounted to a little over $1 billion, including interest, two decades after the spill.

Nonetheless, Exxon claimed that its estimated $2 billion in cleanup costs and $1billion in settlement of other civil and criminal charges should limit punitive damages to $25 million. Ultimately, Exxon recovered a significant amount of its expenses through insurance claims. In a settlement with the federal government, by accepting civil rather than criminal responsibility, Exxon was able to offset a significant portion of their cleanup costs and losses through tax reductions valued at hundreds of millions of dollars.

Public policy response

The public outcry from the accident and its aftermath forced a political response as well. The U.S. Congress passed the Oil Pollution Act (OPA) of 1990 that, among other things, would prevent ships with a history of a significant spill above 1 million gallons from operating in Prince William Sound. This law also mandated a phasing-in of double-hulled tankers by 2015.

OPA also limited the liability for economic damages from a spill to $75 million, so long as the responsible party is not deemed grossly negligent or in violation of any federal safety regulation, by the principal party or any of its agents or contractors, in the release of oil. This liability cap applies only to the economic damages suffered by such entities as property owners, fishermen, tourism operators, and others whose commerce is reduced by the spill. However, the offending party remains fully responsible for the cleanup costs associated with the spill.

While the $75 million economic damages liability cap seems modest now, and even in light of the economic damages imposed on ExxonMobil, it was actually much harder before the act to force such liability. Maritime law at the time actually barred tourism-related claims unless their claims were directly related to waterfront property damage. In addition, OPA imposes strict liability on the transgressor. As a consequence, plaintiffs would no longer have to prove that the violating entity was negligent.

In 1998, Exxon sued the federal government claiming that the provisions of the OPA amounted to a bill of attainder. This legal premise is that Congress cannot pass laws specifically directed at one legal individual.[36] The 9th U.S. Circuit Court of Appeals denied Exxon's appeal in 2002.

The final analysis

Once proceedings cycled through Alaskan courts, the Federal District Court, the U.S. Ninth Circuit Court of Appeals, and the U.S. Supreme Court, Exxon had pled guilty to three criminal counts for environmental damages. While these counts were punishable by fines of up to $3 billion, they were settled for a $25 million, non-tax-deductible criminal fine, $100 million in environmental remediation restitution to the State and Federal entities that took over the cleanup, and a $900 million tax-deductible civil penalty that was payable over ten years.

Exxon also paid $75 million in an unpublicized settlement to some large commercial fisheries entities, under agreement not to sue, and fought a class action suit on behalf of 32,677 commercial fishermen, Native Alaskans, and other entities for economic damages. Exxon successfully argued to the Supreme Court that the various other fines, penalties, and cleanup costs for its environmental damage acted as a sufficient deterrent, and was able to reduce the $5 billion punitive damages awarded by an Alaskan jury to $507.5 million plus interest dating to 1996. In addition, the $900 million settlement for actual economic damages, with discounting and tax write-offs, was determined by the Congressional Research Service to be between $655 million and $716 million.[37]

Against these various amounts, Exxon received insurance proceeds from Lloyds of London of $780 million.

Public relations response

The pictures of thousands of dead, oil covered animals were stark images etched in to the public mind in 1989. A media well versed on visuals that would compel the public to view the story reinforced these dramatic images. Already, CNN had been pioneering constant 24-hour news coverage, and this story remained in the public's eye for a number of reasons.

Obviously, time is of the essence in mitigating the damage from a major oil spill. With boats and helicopters circling, it took Exxon and Alyeska almost half a day to deploy booms around a spill only a handful of miles beyond the Alyeska terminal and response team. Exxon spokesmen were advised early on by their legal team to understate the extent of damage. The executive team did not comment in a substantive manner for six days, and the Chief Executive Officer Lawrence Rawl did not make a trip to Prince William Sound for almost three weeks.

To further compound media frustrations, Exxon carefully chose the venues for interviews, declaring "It was Valdez – or nothing."[38] Exxon realized that Valdez was a company town, with a much lower level of frustration than would be found in the nearby fishing town Cordova that was much more profoundly affected by the spill.

Exxon was also quietly negotiating behind the scenes with the government to avoid criminal sanctions, and settled in secret with a group of significant fishery claimants under a promise not to sue or publicize the terms of the settlement. In essence, Exxon was choosing venues and soliciting interested parties so it could more effectively control the message. At the same time, it blamed federal and state officials for delaying their cleanup, even as it spent $1.8 million to purchase favorable messages in 166 newspapers that offered an apology but did not accept responsibility. And, when asked how it would pay for a spill that would represent perhaps a year's worth of Exxon's corporate profits, an Exxon executive quipped that it would simply raise fuel prices to pay for the incident.[39]

Ironically, the temporary closing of the Alyeska Oil Terminal created a shortage of crude oil for West Coast refineries, raised gasoline and fuel prices substantially, and helped contribute to strong profits for Exxon in the aftermath of the spill.

A safety response

The public is forgiving of an isolated act. However, the cost to Exxon for a repeated act would be severe indeed, and would rip off the bandages of wounds once healed. Exxon knew that it was necessary to engage in a new corporate policy that emphasized safety. Consequently, Exxon's corporate-wide safety program is now state-of-the-art in the industry. For instance, Exxon was recently willing to close up and walk away from a potentially highly profitable well, but also a highly risky well in the Gulf of Mexico.

The Blackbeard West region, 28 miles off the coast in the Gulf of Mexico, was one of the most challenging, but potentially lucrative, reservoirs when Exxon began to drill in 2005. The reservoir was six miles below the seabed, or almost two times deeper, and in deeper water than the Deepwater Horizon spill. At these depths, it represented immense pressures and temperatures. The possibility of over a billion barrels of oil could return to the company more than $150 billion, given the escalated oil prices in 2006 and 2007.

However, as ExxonMobil drilled their exploratory well, pressures started to rise rapidly with the drill head just 2,000 feet from the reservoir. For fear of a blowout, ExxonMobil abandoned its investment of more than $185 million and almost a year and a half of work. Subsequently, another company McMoRan Exploration, continued at the site, and within seven months declared that they had successfully tapped into a reservoir of between half a billion and several billion barrels of oil.

Some analysts declared that Exxon had become timid in an industry long known for taking risks. "They would have done just fine... They just didn't have the guts," proclaimed oil analyst George Froley.[40] Rick Steiner, of the University of Alaska Fairbanks, and a marine biologist and industry critic, commented that, by walking away from a promising prospect, Exxon's recently exhibited caution, is unusual for the industry.[41]

However, strong conclusions from anecdotal evidence are not necessarily sound. Rather, one must look at longer term safety records. For instance, before the BP Deepwater disaster that caused the loss of eleven lives and the release of almost five million barrels of oil, the Deepwater Horizon drilling rig had gone for seven years without a serious injury.

I next turn to the safety records of U.S. oil exploration and production operations.

10
A Brief History of Oil Rig Fires

Gas and oil exploration and extraction is intrinsically dangerous. Electrical and diesel-fuel operated equipment must be used to power machines that move highly flammable liquids and gases. While extensive precautions are taken to minimize the risks, fires and explosions on rigs and in distribution and storage networks are actually relatively common. In fact, fires and explosions in the Gulf of Mexico region occur more than 128 times each year, on average (Table 10.1).

Certainly, no fire hazard is as potentially perilous as one that occurs on an offshore rig. On the larger rigs such as the Deepwater Horizon, more than a hundred people work on a platform that is frequently tens or hundreds of miles from shore and positioned 100 feet or more above the ocean's surface. While rigs are designed to minimize fire danger by isolating certain platform areas from spark and open flame, fires are not uncommon. Indeed, barely five months after the Deepwater Horizon explosion and fire, there was an eerily similar accident on the

Table 10.1 Fires and Explosions in the Gulf of Mexico Oil and Gas Extraction Industry, 2006–2009

Year	2006	2007	2008	2009
Major (greater than $1 million damage)	2	0	0	0
Minor (between $25 thousand and $ 1million damage)	8	6	9	3
Incidental (less than $25 thousand damage)	122	104	130	130
Fire and Explosion Totals	132	110	139	133

Source: http://www.boemre.gov/incidents/firesexplosion.htm, accessed October 25, 2010. Data from the Bureau of Ocean Energy Management, Enforcement, and Regulation.

Gulf production platform Vermilion 380 owned by Mariner Energy, Fortunately, the rig fire that forced the evacuation of 13 crewmembers did not result in any loss of life.[42]

An early lesson learned

Much of what we know of the minimization of risk arose from the Piper Alpha rig tragedy in the North Sea in 1988.

The Piper Alpha was an exploration rig owned by Occidental Petroleum. It was a production platform, meaning that it did not drill for oil, but rather extracted oil, originally, and, subsequently, natural gas. It was operating in the North Sea a little more than 350 nautical miles south of the Arctic Circle. It had tapped into and started producing from the Piper oilfield in 1976 and had been producing upwards of 300 thousand barrels of oil per day. The oil travelled from the production rig to an oil terminal through a thirty inch pipe that was 128 miles long.

The oil formations in the North Sea were some of the leading offshore reservoirs in a deep-sea exploration and production world that also included the Canadian fields off Newfoundland. The United Kingdom and Norway each have learned to stay on the safe side of the reward–risk tradeoff in their oil production, with some exceptions.

For instance, the former platform *Alexander Kielland* had been converted to a floating hotel, or floatel, to service Norwegian workers in their North Sea fields. In March of 1980, the floatel capsized, killing 123 of its 212 workers. On hire to Phillips Petroleum at the time, 40 knot winds and 40 foot waves had battered the rig until a resounding crack was heard from its underwater supports. Despite the fact that the rig would not collapse for another 14 minutes, evacuation was relatively ineffective.

Just eighteen months later, on February 15, 1982, the *Ocean Ranger* rig, off Newfoundland's coast, sank in similar weather conditions, causing the loss of 84 lives. A rogue wave had penetrated a window and initiated flooding of the semisubmersible's ballast. Very much like the loss of life at the *Alexander Kielland*, inadequate evacuation plans caused great loss of life.

Such large scale loss of life would not be the industry's last.

The Piper Alpha platform had been designed as a state-of-the-art oil production rig in 1973. The platform included many safety features, including the separation of crew living quarters from the area of the platform that moved oil from nearby fields. However, the platform had been modified to permit the processing of natural gas as the fields near the rig began to be depleted of oil. Consequently, safety designs optimized for oil production were compromised as the platform was converted to natural gas production.[43]

On July 6 of 1988, nothing was out of the ordinary. Recent work on the platform to route pressurized natural gas had been completed in the previous weeks. One of the pumps used to move natural gas was undergoing maintenance of a duration that could not be completed in a single work shift. As is customary in such instances, the valve to a pipe attached to the pump under repair had been turned off and the pipe sealed with a temporary flange. The required documentation was completed to instruct the next shift to not use the pump in question.

On that fateful day, the engineer responsible for the maintenance plan was unable to meet and describe to the next crew the status of the pump. Instead, he relied on documentation that apparently was never reviewed by the next crew.

Shortly after the beginning of the evening shift, the flow stopped in one of the two pipes of liquid natural gas that powered the platform. Because the platform and all of its emergency systems relied on this supply, the platform manager had to either restart the pump in the line that had just failed or switch to the pump in the line that was undergoing maintenance. Unable to find documentation on the maintenance status of this second system, he switched on the system undergoing maintenance. The resulting pressure in the line blew out the temporary metal sealing flange. Gas began to leak into the pump room at high pressure and triggered a number of gas alarms. However, before the emergency stop switch could be activated, the gas came into contact with heat or spark and ignited.

The shrapnel from the explosion caused a rupture in another gas pipe and created a second source of fire. Within just a few minutes, at 10:04 p.m., with the sky still bright on a summer night not far from the Arctic Circle, the massive Piper Alpha platform initiated its abandonment procedure.

The resulting gas explosion was not of the type that is encountered when crude oil catches fire. Consequently, the fire protection measures built into the platform originally designed for oil was not adequate to protect the platform and its crew from explosive gas.

With the control room abandoned first, coordination of the emergency plan became difficult. Some brave individuals attempted to manually activate the platform's emergency firefighting systems, but to no avail. Others waited for rescue assistance in an area that was considered fireproof. Meanwhile, blown-out pipes maintained a steady supply for an increasingly ferocious fire. Soon, more pipes that fed the gas distribution platform would burst and would engulf the platform in a huge fireball.

Within two hours, the bulk of one of the world's most massive off-shore oil and gas platforms had sunk, along with 165 crewmembers. Only 59 crew members survived. Another two men from an emergency boat died while trying to rescue the crew.

The pipes that had fed the distribution platform continued to burn for another three weeks. Red Adair, the world famous runaway well capper, finally managed to extinguish the flames and stop the flow of gas, despite 80 mph winds and 70 foot seas. Paul "Red" Adair was the same blowout expert who had tamed the Ixtoc I blowout in the Gulf of Mexico in 1979 and the Gulf War spills in 1991, who was immortalized in the John Wayne movie "Hellfighters,"[44] and died of natural causes on August 7, 2004, at the age of 89.[45] Recently, the firm known as Boots and Coots, created by his former employees, Asger "Boots" Hansen, and Ed "Coots" Mathews, completed the bottom kill that secured the runaway Deepwater Horizon spill.

Much was learned from the Piper Alpha catastrophe. The accident created a better understanding of the prevention of migrating gas and of gas explosions. Redundant emergency power systems also became imperative so the failure of one, or even two systems would not induce a hazardous system. We shall discover that redundant systems, and well-designed mechanisms to separate gas and oil from spark and flame can fail, as they did more than two decades later on the Deepwater Horizon rig.

Most significantly, though, was the acknowledgement of the potential for a serious accident should elegant engineering solutions fail to protect a rig and crew from simple human error. As a consequence of these mechanical and human failures, the governments of Canada, Norway, and the United Kingdom determined that risk-based precautions are necessary. For instance, all else equal, a location like those in the North Sea or off of Newfoundland, must be designed to include more fail-safe and precautionary engineering because remoteness makes rapid outside assistance impossible. In addition, as technologies begin to develop riskier fields, further precautions must be taken. With greater risk must inevitably come greater regulation and precaution. This regulatory reality, put in place in these countries, but less so in the United States, rebalances the familiar reward/risk tradeoff previously exercised by drilling engineers. Legislators realized that regulators, as reactive entities, couldn't be expected to anticipate risk. Instead, this responsibility must also be undertaken by the operators who can best assess the risks they undertake.

Indeed, the loss of lives in explosions in the Gulf of Mexico caused the Marine Board of the U.S. National Research Council to recommend the MMS revise regulation to require operators to accept more responsibility for risk. The Board's fear was that a prescriptive regulatory regime, like the regulatory theory the MMS subscribed, created a sense among operators that their adherence to regulatory edict was sufficient to protect their company and their crew. These 1990 recommendations, in the cacophonic aftermath and lobbying of the Exxon Valdez, were never acted upon. Instead, the prescriptive regulatory regime of the MMS was made even more prescriptive.[46] By the time of the Macondo spill, the MMS had still not developed a new risk management protocol, even though two decades had elapsed since the board recommendations.[47] Indeed, over the two decades since the Exxon Valdez disaster, the MMS budget for leasing, environmental safeguards, and regulation remained constant, in real terms, while offshore oil production quadrupled and the real value of oil production went up 25 fold.

11
Exploration, Drilling, and Extraction U.S. Environmental and Safety Records

The oil exploration and drilling industry has long maintained a reputation of roughnecks, John Wayne-type rugged individuals swiftly executing dangerous procedures in search of a volatile and explosive liquid contained in pressurized pockets beneath the earth's surface. When things, go wrong, as they occasionally do, a daring Red Adair-type individual sweeps in to explode the runaway well back into the earth.[48]

This risk-taking reputation for the industry has mostly been replaced by a new cadre of thoughtful and well-schooled geological engineers and physicists who now exploit some of the world's most sophisticated technologies in search for oil increasingly more difficult and more dangerous to extract.

However, these new technologies and newly trained experts are expensive. Only the largest and best capitalized companies can afford the most sophisticated engineers, oil drilling platforms, seismic sensing, and computer-assisted subterranean mapping. The culture of rugged, risk-taking roughnecks are now mostly relegated to myriad small, family-run drilling contractors, or some larger contractors willing to take on greater risk for returns.

Certainly, there are other aspects of the fossil fuels industry that remain risky. On September 9, 2010, a little more than four months after the Deepwater Horizon spill, eight people lost their lives when a gas line exploded and leveled a neighborhood of San Francisco, California. When it was subsequently discovered that there were complaints from residents of the characteristic odor mixed with otherwise odorless natural gas that had gone unheeded, Pacific Gas and Electric Company was forced to revise their safety plans.[49] It was subsequently reported that the installation of a simple remote shutoff valve could have ameliorated the damage and saved eight lives.

Even the largest companies are prone to refinery risks, given the proximity to calamity when technologies depend on heating volatile and explosive hydrocarbons and combining them with other reactive chemicals in an environment that measures volumes by the millions of gallons.

For instance, a blast and fire at a Tesoro-owned refinery in Anacortes, Washington, killed four people in April of 2010, the same month of the BP Deepwater Horizon disaster. Six months earlier, a flare stack overflowed at a Tesoro-owned refinery in Salt Lake City, Utah, causing a fire and explosion.

A little more than four years before that, in March of 2005, fifteen people were killed at a Texas City, Texas BP refinery explosion, caused when an octane-boosting unit overflowed as it was being restarted. BP quickly realized it had inherited a problem refinery when it took over the Amoco facility as part of its merger with the U.S. company. While this merger in 1998 with Amoco created an instant expansion of BP's U.S. presence, it also created new risks for BP.

Amoco had a long and unfortunate history in the petroleum industry before its merger with BP. It was the responsible party for one of the first major supertanker spills in a populated and ecologically sensitive area. Indeed, the Amoco Cadiz spill did much to subsequently ratchet up the expectation for oil spill readiness.

Just two years after the Amoco Cadiz spill of 1978, an explosion of Amoco's chemical plant in New Castle, Delaware, killed six people and caused the loss of 300 jobs.[50] A decade after that, two people died in a refinery explosion at Amoco's Whiting refinery in Indiana, near Chicago. The third largest oil refinery in the United States, this refinery suffered another fire a little more than five years after Amoco merged with BP, injuring three people.

Of course, refinery dangers are not limited to Amoco, or its more recent parent company, BP. As recently as 2007, a fire rocked the Chevron Pascagoula, Mississippi refinery, and, in a six-month period alone, fire and explosions rocked Exxon refineries in Baton Rouge, Louisiana, Trecate, Italy, and Fawley, England, and Chevron refineries in Corio, Australia, and Anacortes, Washington.

An exploration of Occupational Safety and Health Administration (OSHA) violations or Environmental Protection Agency violations of air quality by refineries located in the United States would demonstrate that oil companies often have checkered relationships with state and local regulators.

Such an analysis is relatively meaningless for this case study, though, because refining and distribution are operations distinct from exploration, employ a very different set of technologies, and are operated within a significantly different corporate subculture. Instead, we will next explore the much more relevant OSHA health and safety records and the U.S. Environmental Protection Agency (EPA) records of environmental violations by major oil exploration and drilling companies.

EPA violations in the oil exploration industry

The EPA creates reports for incidents of air or water pollution among oil exploration companies based on inspections, citizen reports, worker reports, or self-reports by the companies themselves. These incident reports can be viewed by the public and have been digitally documented since September, 2003.

An analysis of these reports for all companies engaged in exploration and drilling in the United States demonstrates that oil companies are frequently subject to numerous violations. Table 11.1 shows that Chevron was the subject of 1,269 incident reports, BP was invoked as the responsible party in 1,206 reports, ExxonMobil in 976 reports, and Shell in 464 incident reports.

Interestingly, while these four companies represent the majority of market share in U.S. oil drilling and exploration, the records of much smaller companies are much spottier, when their smaller size and scale of operations is considered.

For instance, Taylor Energy, a privately owned independent driller active in the Gulf of Mexico, was responsible for 860 incident reports. Moreover, Apache Oil, a Houston, Texas oil company with annual revenues of US$8.615 billion in 2009, or 2.8% the size of ExxonMobil, had 371 incident reports over the period, or about 38% of the number of incidents reported in which ExxonMobil was identified as the responsible party.

Indeed, Mariner Energy, another Texas-based oil exploration company that suffered a Gulf of Mexico platform explosion on September 2, 2010, in the aftermath of the BP Deepwater Horizon fire, had 100 violations since 2003.

It is clear that all the large oil companies have proportionately fewer environmental incidents than the average in their industries. The large companies can afford, and are dedicated to, more elaborate systems and technologies that, while imperfect, result in fewer incidents than the smaller companies in the industry.

OSHA violations by oil drilling companies

The larger companies also have more elaborate safety systems than their smaller counterparts. The U.S. Department of Labor's OSHA lists data

Table 11.1   EPA violations by oil exploration companies since September 2003

Suspected responsible company	Number of incident reports to EPA, September 2003–January 2010
Chevron	1269
BP	1206
Exxon/Mobil	976
Taylor Energy	860
Shell	464
Apache	371
Energy Partners	211
Wild Well Control	180
Stone Energy	158
Plaines Exploration and Production	146
Diamond Offshore	130
Dcor	112
Unocal	109
Maritech Resources	107
Helis Oil & Gas	103
New Field Exploration	102
Mariner Energy	100
W & T Offshore	97
Devon Energy	88
El Paso E&P Company LP	84
SPN Resources	75
Conoco Phillips	68
Freeport Mcmoran Energy	66
Anadarka	60
Kerr Mcgee	60
Marathon Oil	60
Energy Resource Tech	58
Cox Operating	53
Venoco Inc.	52
Noble Energy	51
Forest Oil	50
Energy Xxi Llc	49
Merit Energy	45
Atp Oil & Gas	44
Bois D Arc	43
Swift Energy	42
Mcmoran Oil & Gas Llc	41
Aera Energy	39
Hillcorp Energy	39
Marlin Energy	37
Murphy Exploration	36
Nexen Petroleum	35
Hunt Oil Co	34
Arena Offshore	30
Pogo Producing	30
Total Oil Company	30

Source: Constructed from Environmental Protection Agency incident reports published at www.epa-echo.gov/echo/

for routine health and safety inspections, the results of accident-related inspections, and violations arising from complaints.[51]

While OSHA records violations and fines for all industries, I focus here on the safety records for drilling operations. The data is organized around Standard Industrial Classification codes, of which 1381 is the SIC code for the industry engaged in drilling oil and gas wells. These are:

> Establishments primarily engaged in drilling wells for oil or gas field operations for others on a contract or fee basis. This industry includes contractors that specialize in spudding in, drilling in, redrilling, and directional drilling (perform the following operations):
> - Directional drilling of oil and gas wells on a contract basis
> - Redrilling oil and gas wells on a contract basis
> - Reworking oil and gas wells on a contract basis
> - Spudding in oil and gas wells on a contract basis[52]

There are a handful of corporations that dominate the OSHA safety violations record over the decade 2000–2009. Note that none of the large firms of ExxonMobil, Chevron, Shell, BP, ConocoPhillips, or Texaco are included in the list of drilling companies with 20 or more safety violations over the decade. Over that period, BP's drilling and exploration operation had four violations, and was fined $6,750 as a consequence of a routine inspection in Alaska in 2006. Shell Oil had three violations and a fine of $4,500 resulting from a routine inspection of a facility in Texas, also in 2006.

While drilling-related accidents and oil discharge statistics suggest that large companies are under-represented in the statistics and run safer and cleaner operations than the industry average, the sheer potential for damage from the largest deep-sea drilling activities is obviously large. The BP Deepwater Horizon accident attests to the peril a calamity on one large scale operation can produce.

The relatively small number of very large deep-sea platforms implies that accidents should be quite rare. Each oil company maintains a safety protocol that attempts to minimize the risks we will explore in the BP Deepwater Horizon disaster. Moreover, the protocol of one drilling company may determine a circumstance presents unacceptable risk, another driller may consider the risk manageable. Recall ExxonMobil's decision to abandon drilling at its lease in the Blackbeard West region in 2006. The company had drilled to within 2,000 feet of a reservoir six miles beneath the sea floor. Rising pressures exceeded ExxonMobil's acceptable risk protocol. However, McMoRan Exploration found the risk manageable, purchased the lease from ExxonMobil, and successfully

Table 11.2 OSHA violations by oil drilling companies

Corporation	OSHA violations
Patterson-UTI Drilling Company Llc	302
Nabors Drilling USALP	225
Cdx Resources LLC	167
Cyclone Drilling, Inc.	133
Grey Wolf Drilling Company LP	129
Unit Drilling Company	122
Helmerich & Payne International Drilling Co	112
SST Energy Corporation	103
DHS Drilling Company	99
Union Drilling Inc.	77
Ensign United States Drilling Inc.	67
True Drilling LLC	65
Gene D. Yost & Son, Inc.	55
Pioneer Drilling Services Ltd	46
Capstar Drilling, LP	37
Top Drilling Corporation	37
Scorpion Exploration & Production Inc.	36
Corpus Christi Drilling & Workover L.L.C.	35
Warren Drilling Company, Inc.	35
Caza Drilling Inc.	33
Elenburg Exploration Company Inc.	32
Falcon Drilling Company, LLC.	32
Lariat Services, Inc.	32
Bronco Drilling Company, Inc.	30
Nucor Drilling Inc.	30
Bigard & Huggard Drilling Inc.	28
Precision Drilling Oilfield Services Corp	28
S.W. Jack Drilling Company	28
Premium Well Drilling	25
Price Drilling Company, Inc. Rig #1	25
GWDC America, Inc.	24
Sauer Drilling Company	24
West Texas Energy Services, LLC	24
Rainbow Drilling LLC	23
Trinidad Drilling Ltd	23
Cactus Drilling Company, LLC	22
Kilbarger Construction Inc.	22
Appalachian Drilling, LLC	21
Phoenix Drilling, Inc.	21
Robinson Drilling Of Texas, Ltd	21
Xtreme Coil Drilling Corporation	21
Berentz Drilling Company, Inc.	20
L & M Drilling	20
Nexus Drilling Corporation, Inc.	20
Pinpoint Drilling & Directional Service LLC	20
Texas Wyoming Drilling Inc.	20

Source: http://www.osha.gov/pls/imis/industry.html, last accessed March 23, 2011.

tapped into a reservoir it estimated to contain between half a billion and several billion barrels of oil.

Oil releases from 1964 to 2009

12158 barrels of oil were spilled from offshore sites in the U.S., primarily into the Gulf of Mexico, annually on average over the 45 years from 1964 to 2009. Over the last ten years from 2000 to 2009, the accidental discharge rate fell to an annual average of 6,002 barrels. BP Exploration spilled 777 barrels annually on average over this period, or 12.9% of the annual accidental discharge.

To place Gulf of Mexico oil production in perspective, the top ten exploration companies produced 284.8 million barrels of oil in the Gulf of Mexico in 2009, according to the Bureau of Ocean Energy Management, Regulation, and Enforcement, out of a total of 356.4 million barrels by all Gulf of Mexico operators. The top three producers, BP Exploration, Shell Offshore, and Chevron USA, produced 30.1%, 23.1%, and 7.6%, respectively, of the crude oil extracted from the Gulf of Mexico in 2009. These three largest companies accidentally discharged one barrel of oil, on average, for every 159,965 barrels extracted. BP Exploration was just under the Big Three average, with one barrel of oil spilled for every 138,278 barrels extracted. However,

Table 11.3   Offshore oil production quantities among the various major oil companies

Rank	Operator	Crude Oil (BBLS)
1	BP Exploration & Production Inc.	107,442,475
2	Shell Offshore Inc.	82,356,357
3	Chevron U.S.A. Inc.	27,115,565
4	Kerr-McGee Oil & Gas Corporation	21,484,973
5	Exxon Mobil Corporation	9404,313
6	Apache Corporation	12,371,503
7	Eni Petroleum Co. Inc.	12,086,274
8	Hess Corporation	1,478,480
9	BHP Billiton Petroleum (GOM) Inc.	5,222,995
10	Energy XXI GOM, LLC	5,838,139
	Subtotal	284,801,074

Source: According to the Bureau of Ocean Energy Management, Regulation, and Enforcement. http://www.boemre.gov/stats/PDFs/RankbyOperator072909.pdf, accessed September 4, 2010.

Table 11.4   Offshore lease activity

Year	Well drilling status			Completed (Wells and boreholes)	Plugged & abandoned (Wells and boreholes)	Cumulative total
	Active	Suspended	Other[1]			
1960	58	53	N/A	1,923	686	2,720
1961	54	99	N/A	2,467	814	3,434
1962	56	107	N/A	3,091	1,002	4,256
1963	62	130	N/A	3,631	1,226	5,049
1964	73	193	102	4,313	1,372	6,053
1965	89	261	0	4,733	1,685	6,768
1966	82	0	444	3,305	1,871	5,702
1967	95	0	496	3,762	2,223	6,586
1968	133	0	592	4,258	2,592	7,575
1969	103	129	590	4,752	2,919	8,493
1970	106	115	534	5,359	3,278	9,392
1971	89	152	551	5,718	3,724	10,234
1972	79	263	537	6,032	4,168	11,079
1973	84	249	546	6,421	4,599	11,899
1974	91	292	1,006	6,218	5,108	12,715
1975	97	292	977	6,104	5,617	13,087
1976	97	362	1,117	6,461	6,088	14,125
1977	169	496	N/A	7,914	6,610	15,189
1978	117	456	N/A	8,433	7,133	16,169
1979	175	603	N/A	8,954	7,576	17,318
1980	191	739	N/A	9,638	8,057	18,625
1981	173	724	N/A	10,308	8,704	19,909
1982	166	701	N/A	11,164	8,913	20,944
1983	134	597	N/A	11,990	9,374	22,095
1984	253	313	1,151	11,861	9,903	23,481
1985	195	346	1,166	12,285	10,487	24,481
1986	95	279	1,200	12,536	11,909	26,019
1987	142	265	1,275	12,736	12,373	26,791
1988	116	289	1,402	12,827	13,164	27,798
1989	123	361	1,441	12,938	13,846	28,709
1990	120	266	1,466	13,167	14,677	29,696
1991	64	249	1,436	13,184	15,430	30,363
1992	104	150	1,455	13,209	16,348	31,306
1993	129	193	1,433	13,181	16,709	31,645
1994	117	222	1,435	12,705	16,860	31,339
1995	124	247	1,522	13,475	18,089	33,457
1996	212	151	1,615	13,583	18,817	34,378
1997	268	149	1,792	13,546	19,956	35,711

Continued

Table 11.4   Continued

Year	Well drilling status			Completed (Wells and boreholes)	Plugged & abandoned (Wells and boreholes)	Cumulative total
	Active	Suspended	Other[1]			
1998	175	122	1,913	13,702	21,124	37,036
1999	219	110	2,206	13,011	22,034	37,400
2000	230	146	2,166	13,096	22,735	38,373
2001	153	73	2,032	13,930	24,474	40,662
2002	143	73	2,116	13,876	25,484	41,692
2003	204	50	3,138	18,424	32,251	54,067
2004	197	58	3,296	18,260	33,746	55,557
2005	242	67	3,601	18,001	34,878	56,789
2006	209	61	3,834	17,601	36,470	58,375

Note: [1] http://www.boemre.gov/stats/PDFs/OCSDrilling.pdf, accessed September 4, 2010.
Source: According to the Bureau of Ocean Energy Management, Regulation, and Enforcement.

Table 11.5   Offshore oil spills by company 1964–2009

Company name	Number of incidents 1964–2009	Offshore spills (bbls)
Humble Oil Co.	2	160,866
Shell Offshore, Inc.	40	81,507
Union Oil Company of California	6	81,299
Chevron U.S.A. Inc.	36	58,531
Pennzoil Company	6	25,256
Amoco	12	18,743
Signal Oil & Gas Company	2	15,135
Exxon Mobil Corporation	8	11,094
BP Exploration & Production Inc.	23	8,639
Gulf Oil Corporation/Chevron	4	6,594
Atlantic Richfield Company	5	5,433
Midwest Oil Corp. &/or Continental Oil Co.	1	5,180
Total E&P USA, Inc.	2	4,917
Texaco, Inc.	7	4,038
BHP Petroleum Company Inc.	4	3,540
Continental Oil Company	7	3,433
Seashell Pipeline Company	1	3,200

Continued

Table 11.5   Continued

Company name	Number of incidents 1964–2009	Offshore spills (bbls)
Mobil Oil Exploration & Producing Southeast	7	3,126
Taylor Energy Company	3	2,416
Mariner Energy, Inc.	5	2,323
Equilon Pipeline Company LLC	1	2,240
Forest Oil Corporation	3	2,216
Anadarko Petroleum Corporation	6	2,173
Pan American Petroleum Corp.	1	1,688
Murphy Exploration & Production Co.	6	1,642
Tenneco Oil Company	1	1,589
Texoma Production Company	2	1,580
Remington Oil and Gas Corporation	1	1,572
Hunt Petroleum (AEC), Inc.	1	1,494
Noble Energy, Inc.	8	1,475
W & T Offshore, Inc.	5	1,351
Marathon Oil Company	3	1,347
High Island Offshore System, LLC	1	1,316
Kerr-McGee Oil & Gas Corporation	4	1,263
Stone Energy Corp. (SEC) former Bois d'Arc	6	1,193
Apache Corporation	7	1,119

Source: According to the Bureau of Ocean Energy Management, Regulation, and Enforcement.

smaller operators often have substantially poorer records. The average for the ninth ranked offshore oil producer, BHP Billiton Petroleum, was one barrel of oil accidentally discharged for every 14,754 barrels produced.

Safety records

Data from the Bureau of Ocean Energy Management, Regulation, and Enforcement, (formerly called the Minerals Management Service, or MMS) the group that oversees and tabulates regulatory compliance of offshore energy producers, shows that similar safety conclusions can be drawn over representative injury rates (Table 11.7).[53]

We see that, while the three largest companies, BP Exploration, Shell Offshore, and Chevron, represent 60.8% of crude oil extracted in the

Table 11.6   Oil exploration company spills 2000–2009

Company name	Number of incidents 2000–2009	Number of barrels spilled
BP Exploration & Production, Inc.	22	7,769
Total E&P USA, Inc.	2	4,917
BHP Petroleum Company Inc.	4	3,540
Chevron U.S.A. Inc.	18	3,343
Shell Offshore, Inc.	17	2,447
Taylor Energy Company	3	2,416
Mariner Energy, Inc.	5	2,323
Equilon Pipeline Company LLC	1	2,240
Forest Oil Corporation	3	2,216
Anadarko Petroleum Corporation	5	2,107
ATP Oil & Gas Corporation	1	1,718
Murphy Exploration & Production Company	6	1,642
Remington Oil and Gas Corporation	1	1,572
Hunt Petroleum (AEC), Inc.	1	1,494
Noble Energy, Inc.	8	1,475
W & T Offshore, Inc.	5	1,351
High Island Offshore System, LLC	1	1,316
Stone Energy Corp. (SEC) former Bois d'Arc	6	1,193
Apache Corporation	7	1,119
LLOG Exploration Offshore, Inc.	3	930

Source: According to the Bureau of Ocean Energy Management, Regulation, and Enforcement.

Gulf of Mexico in 2009, they represented just 22 of the 130 injury reports in 2009. Overall, the large producers have relatively better injury records than industry averages.

Fatalities statistics

This safety record among the largest deep water drilling companies is also reflected in the annual Minerals Management Service fatalities reports, published online since 2006. Over the four-year period from 2006 to 2009, there were a total of 30 fatalities. With the exception of Chevron, which was the listed responsible party for three fatalities in

Table 11.7 Worker injuries in exploration by major oil exploration companies

Company name	Number of injuries 2009
Apache Corporation	11
Chevron U.S.A. Inc.	9
Murphy Exploration & Production Company – USA	9
Mariner Energy, Inc.	9
BP Exploration & Production Inc.	8
W & T Offshore, Inc.	7
Wild Well Control, Inc.	6
Shell Offshore Inc.	5
Kerr-McGee Oil & Gas Corporation	5
ATP Oil & Gas Corporation	4
ENI US Operating Co. Inc.	4
Energy XXI GOM, LLC	4
DCOR, L.L.C.	3
El Paso E&P Company, L.P.	3
Anadarko Petroleum Corporation	3
Union Oil Company of California	3
Exxon Mobil Corporation	3
Hunt Oil Company	2
Energy Partners, Ltd.	2
Energy Resource Technology GOM, Inc.	2

Source: According to the Bureau of Ocean Energy Management, Regulation, and Enforcement.

three separate incidents, and Exxon, with one fatality, all U.S. fatalities occurred at platforms owned by companies other than the major oil companies (see Table 11.8).

USA Today recently reported that Exxon paid $205,000 in fines for safety and other violations in the Gulf of Mexico for the period 1997–2010. Over the same period, BP paid $616,000 in fines.[54] However, BP's presence in the Gulf of Mexico is by far the largest of all major producers, with a production level more than eleven times that of ExxonMobil.

BP's safety record was called into question following the Deepwater Horizon spill. From data that goes back to 1999, the year Amoco, a company with a more notorious safety and spill record, was fully integrated into BP, the record of safety at BP was worse than in recent years. If the years 1999–2009 are averaged together, one could conclude that BP's record, across the corporation, was not as good as some other major oil companies. However, BP has demonstrated significant safety

Table 11.8   Four year fatality figures for oil exploration companies 2006–2009

Responsible party	Fatalities 2006–2009
ANR Pipeline	3
Chevron U.S.A. Inc.	3
Energy Resource Technology, Inc.	2
McMoRan Oil & Gas LLC	2
Remington Oil and Gas Corporation	2
Anadarko Petroleum Corporation	1
Apache Corporation	1
ATP Oil & Gas Corporation	1
Badger Oil Corporation	1
Devon Energy Operating Company, L.P.	1
El Paso Production Oil & Gas Company	1
Enterprise Field Services, LLC	1
Exxon Mobil Corporation	1
Forest Oil Corporation	1
Freeport-McMoRan Energy LLC	1
Hall-Houston Exploration II, L.P.	1
Linder Oil Company, A Partnership	1
LLOG Exploration Offshore, Inc.	1
Newfield Exploration Company	1
Peregrine Oil & Gas II, LLC	1
Repsol E&P USA Inc.	1
Stingray Pipeline Company	1
Stone Energy	1
Total:	30

Source: According to the Bureau of Ocean Energy Management, Regulation, and Enforcement.

enhancements since the BP–Amoco merger. Their "days away from work case frequency" was high just following the AMOCO–BP merger in the 1990s, but fell by between one half and three quarters a decade later. Clearly, BP's absorption of AMOCO caused it to inherit problems that would take a decade to remedy.[55]

The largest drilling companies not only mirror each other in their relatively strong safety and better-than-industry-average environmental records, though. Congressional hearings following the BP Deepwater Horizon spill revealed that they also shared almost identical emergency response plans. At a House Energy Committee hearing on June 15, 2010, executives from ExxonMobil, ConocoPhillips, Chevron, and Royal Dutch Shell attempted to distance themselves from their bigger brother in Gulf of Mexico oil exploration and extraction. However, law-makers noted that their oil disaster response plans were almost identical to those used by BP Exploration. The same consulting firm had been employed by all five companies to prepare their plans, and often used identical language in all five, including a common reference to walruses in the plan, despite the fact that walruses have never inhabited the Gulf of Mexico. Henry Waxman the committee chair observed "the covers of the five response plans are different colors, but the content is 90 percent identical."[56]

The commonality of practices among the big five companies induced Bart T. Stupak, a Michigan congressman, to conclude that: "It could be said that BP is the one bad apple in the bunch, but unfortunately, they appear to have plenty of company." The chairman of the committee, Congressman Waxman, added: "BP failed miserably when confronted with a real leak, and ExxonMobil and the other companies would do no better."

Indeed, even the executives of other oil companies now seem willing to defend practices overall in the deep-sea exploration industry. Andrew Swiger, Senior Vice President of ExxonMobil recently noted that "[t]he industry has drilled thousands of wells in all manner of operating environments, whether it's deepwater or onshore, all around the world, without incident."[57]

Nonetheless, given what the world now understands as the dangers associated with even one major deep-sea well failure, even by corporations with strong safety records and few prior catastrophic incidents, most would agree that the universal adoption of industry best practices, and strong oversight, are prudent responses to the BP Deepwater Horizon spill.

Part III

The Macondo Prospect and What Went Wrong

The Macondo Prospect, a potentially huge reservoir in the Mississippi Canyon of the Gulf of Mexico, would not give up its riches easily. The prospect, the accident, lessons, and the technical and management challenges are documented here as companies are forced to take greater risks to quench our thirst for oil.

12
The Macondo Prospect

The Latin American novelist Gabriel José de la Concordia Garcia Márquezwon the Nobel Prize for Literature in 1982 for his novels and short stories that often depicted a fictional village in Colombia called Macondo. This village portrayed a sense of magic and solitude. The town's association with magical realism in the novels led some in Latin America to refer to the evolution from solitude to wealth, and back to solitude, or to absurd news events, as belonging to Macondo.

The opportunity to name a potentially rich oil prospect in the Gulf of Mexico was a prize in a charitable auction. The winner named what would become the largest deep water well blowout in history, the Macondo Prospect. The location is an offshore oil lease block located at a latitude of 28.736667 degrees North, and 88.386944 degrees West in the U.S. Exclusive Economic Zone off the coast of Louisiana. It is in the Mississippi Canyon, block 252, and is leased by BP Exploration (65% owner), Anadarko (25% owner), and Mitsui Offshore Exploration 2007 (10% owner). Its lease designation, for short, is MC252.

The site was initially surveyed in 1998, and was the subject of further high resolution 3D seismic surveying in 2003. BP Exploration purchased the mineral rights to the site in a lease sale by the Minerals Management Service (MMS) in 2008.[58] Upon securing the lease, BP conducted additional mapping and subsequently secured approval for a drill plan with the MMS in 2009.

Certainly the Macondo site was not the first challenging well in the Gulf of Mexico. There had been interest in Gulf of Mexico deep-sea drilling, but costs remained prohibitive and challenges daunting. However, rising oil prices in the 1980s and a 1987 Reagan-era decision by the U.S. Minerals Management Service to slash by 85% the minimum bid price on deep-sea leases in the Gulf helped spur renewed exploration.

Almost immediately, after oil prices began to rise in the 1980s, Shell hit it big, but kept their discovery under wraps. Their commissioned Zane Barnes exploratory platform found oil almost 3,000 feet deep and 17,000 feet below the ocean floor. Shell had hit it big earlier in their Bullwinkle well, and had pioneered new techniques that earned them relatively high oil extraction rates in some challenging geological conditions. Their new Auger prospect would redefine oil riches in the Gulf. However, with the high potential for huge fields came much higher costs. Shell would bring in BP as a partner. They could not imagine that, within little more than a decade, BP, then a beleaguered former oil power that lost its positions in the Middle East and in Africa, would emerge as the biggest operator in the Gulf of Mexico.

While BP built up its leases and its expertise over the 1990s, Shell would continue to be the most successful large field developer in the Gulf. Their large Auger field, as a result of the Zane Barnes exploration, was producing over 100,000 barrels of oil and 400 million cubic feet of gas per day by the late 1990s.[59]

Shell's success did not come without regulatory failure, though. When the stakes are large, so are the fines. Shell simply could not keep up with transporting all the gas produced at the field, and was venting incidental gas illegally for four years until the MMS discovered the transgression. Shell ultimately settled for a civil fine of $49 million, which represented just two weeks' production from the profitable Auger field.[60]

Shell had pioneered a new exploratory platform, the tension leg platform, that could drill to depths over a mile. However, as the depths increase, the weight of a number of cables used to anchor the platform to the ocean floor became problematic. Consequently, while Shell was investing in many such platforms, BP found greater success at greater depths with a new type of semisubmersible platform that could hold its position with powerful thrusters and sophisticated satellite-guided navigation and computer systems. Technology would allow even greater complexity in well design and depth, but at the cost of even more complex systems.

BP was also becoming the world leader in a new form of oil well geology. It would be this innovation that would move BP from a struggling oil company to the fourth largest company in the world. The bottom of the Gulf of Mexico has huge and thick salt domes, at great ocean depths. These depths, and the engineering difficulty of imaging and then drilling for oil in salt was problematic. First, of course, was the water depth, and the depth of salt. However, as materials sciences and engineering conquered these challenges, it was the development of new sensing and imaging techniques, and greater computer power, that would allow BP to discover the oil pockets below these salt deposits. This high risk, but high reward financial strategy soon paid off to make BP the dominant upcoming oil company by the mid-2000s.

BPs success was in their ability to undertake numerous deepwater field explorations at a time. Their salt-deposit wells included Mad Dog and Atlantis in the Gulf's Green Canyon, and Thunder Horse in the Mississippi Canyon, the same undersea canyon that holds the Macondo reservoir.

These platforms and wells are also in one of the world's most hurricane-prone areas. Indeed, Mad Dog lost its oil derrick in Hurricane Ike in 2008, and Thunder Horse almost toppled completely, due to improperly installed valves, in Hurricane Dennis in 2005. However, technological and natural hazards were well worth the risk when rewards, as in the case of Thunder Horse, could amount to more than 250,000 barrels of oil per day, or almost 5% of total U.S. production.[61] It was the shear profitability of such prospects, combined with a desperate need for the U.S. to provide for domestic oil as the price of oil approached $150 a barrel in 2008, that induced President George Bush to end the ban on offshore leasing on the continental shelf in 2008, and for President Obama to state:

> Now, here's the last thing I'll say about drilling, though, because what you have is, you have some environmentalists who just said, don't drill anywhere; and then you've got some of my friends on the Republican side who were saying, well, this is a nice first step but it's not enough – you should open up everything.
>
> I don't agree with the notion that we shouldn't do anything. *It turns out, by the way, that oil rigs today generally don't cause spills. They are technologically very advanced* (emphasis added). Even during Katrina, the spills didn't come from the oil rigs, they came from the refineries onshore.
>
> But the notion that we could drill our way out of the problem – you'll start hearing about this because you know what happens during the summer. As soon as gas prices start going up – every summer it's the same thing, right? And then politicians start standing up and – "we're going to do something about it" – and these days some of my colleagues on the Republican side, what they'll say is, you got to drill even more.
>
> Just remember the statistics when you start hearing this. We account for 2 percent of the world's oil reserves but we use 20 percent of the world's oil. We use 20 percent; we only got 2 percent. We can't drill our way out of the problem.

That's why we've got to get moving on this clean energy sector, but we also have to make sure that we've got enough supply that's regular in terms of these other energy – traditional energy, sources, so that by the time we get to the clean energy sector, we haven't had to sacrifice economic growth along the way.[62]

That press release was issued on April 2, 2010, just weeks before the largest offshore oil spill in U.S. history.

The Macondo challenges

BP's strength was in exploration techniques, corporate strategies, and a vertically integrated corporate structure in the United States, much of which was a consequence of its merger with AMOCO. BP did not do its own drilling, cementing, or other more highly specialized duties, though. Instead, BP contracted with companies such as Halliburton for cementing and other oil field services, Schlumberger for such activities as well cement logging, and, especially, Transocean, for the actual exploratory and production drilling.

Such contracting out of the drilling activity is not unusual. While oil companies may on occasion own and operate their own rigs, it is customary to also contract for rigs and crews that are owned and operated by a few highly specialized drilling companies. The highly sophisticated rigs necessary for extreme deep water drilling can cost hundreds of millions of dollars and require extensive expertise to staff and operate. A single oil company, even a major company, will not typically own a sufficient fleet of such specialized equipment because its drilling prospects at a given time may not be expansive enough to fully employ its fleet. In such a case, a company would be reticent to lease its equipment and staff to a competitor, but would be obliged to pay the operating and overhead costs of a very expensive asset.

On the other hand, a drilling company may find itself with more drilling prospects than it can meet with a self-owned fleet. In such a case, the company will find it prudent to lease the equipment and staff of these highly specialized rigs.

Consequently, the industry thrives of companies that specialize solely in drilling. The biggest oil drilling contractor is Transocean, based in Switzerland. One of its rigs, the Deepwater Horizon, was also known as one of the world's most state-of-the-art and sophisticated deep water drilling platforms. Owned, operated, and staffed by Transocean, the Deepwater Horizon rig was leased out at a rate typically in the range of $500,000 per day, including platform workers.

In the case of the fateful Macondo Prospect, BP contracted with Transocean in 2009 to drill the exploratory Macondo well in block MC252. Having negotiated a rate for another Transocean platform and crew, the Marianas semisubmersible rig was moved to the site and began to drill on October 7, 2009. However, that rig was damaged by Hurricane Ida on November 28, 2009. In turn, BP leased a second Transocean semisubmersible rig, the Deepwater Horizon, to recommence drilling. The replacement Deepwater Horizon rig began drilling in February, 2010.[63]

Offshore drilling methods

There are a variety of subsea drilling rigs employed today. The first type is a simple barge or ship from which drilling and piping proceeds below. Such a floating production system (FP) drilling platform is prone to wave action, but it is relatively low cost and effective in shallow waters to a depth of 6,000 feet.

A common type of rig in shallow waters is the fixed platform jackup rig. These designs can be floated to the drill site, and then the legs can be extended to the bottom of the sea. These shallow water fixed platforms that are secured to the seafloor very much resemble their land-based drilling cousins. Because a structurally constrained length

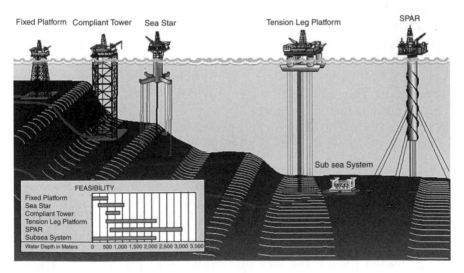

Figure 12.1 Types of offshore drilling platforms
Source: courtesy of the Minerals Management Service.

allows them to only be used at relatively shallow depths, they can be serviced above by drillers using techniques virtually identical to land-based roughnecks, but can also be serviced by divers below. They can be used to a depth of 1,500 feet.

A compliant tower (CT) drill platform is an extension of fixed platforms that can go deeper. Their stability relies on the flexibility of their legs, which allow them to absorb some of the movement of the sea. They can be used in waters deeper than a fixed platform will allow, up to 3,000 feet deep.

Deeper waters yet can employ seastar (SStar) platforms or fixed tension platforms. These floating rigs have a lower hull that is filled with water once it is brought to location. This lower hull is underwater, and hence is somewhat immune to the wave action at the surface. Legs that extend to the surface and secured in place can then be kept under tension to align the platform above the surface. They can be employed in depths upward of 7,000 feet.

SPAR platforms and semisubmersible platforms can drill anywhere from 1,500 feet to 10,000 feet deep. The SPAR platform use a large cylinder hundreds of feet long and nearly 100 feet in diameter that extends from and supports a platform above, and extends toward, but not all the way to, the sea floor. It is tethered in place with a series of cables. Its cylinder is partially filled with water, creating a subsea mass that provides for high stability.

The state-of-the-art platform, in stability but also in versatility, is the design of the Deepwater Horizon rig. The modern semisubmersible rig has a lower hull that can be evacuated of water for moving and filled with seawater to allow the rig to settle into the ocean. The rig does not extend to the sea floor, but rather floats at the surface, suspended by a massive hull below the water line that is protected from sea surface wave action. The mass below water provides for a great deal of stability. This stability is further reinforced by large anchors that secure the ship to the sea floor. This design can operate in very turbulent waters and can also drill deeper than most all other designs.

Drilling in the Mississippi Canyon

The force of the Mississippi River, over hundreds of thousands of years, has carved a canyon deep into the floor of the Gulf of Mexico. Consequently, while there is a fair amount of shallow-water drilling in the Louisiana region, the area protruding from the Mississippi River, along the Mississippi Canyon, is deep and difficult to drill.

The Mississippi Canyon requires a semisubmersible drilling platform that can accommodate the depths encountered at the Macondo Prospect. BP commissioned the Deepwater Horizon drilling rig, a nearly state-of-the-art rig that had been built eight years earlier by the Hyundai Corporation in South Korea, at a cost of $560 million. The rig was rated to weigh up to 32,588 gross tonnes, and could displace 52,587 tonnes of water. It was powered by diesel engines that could provide almost 60,000 horsepower, and was able to cruise on its own power at 4 knots.

The vessel, with a crew of upwards of 146 people, was able to use its thrust system to dynamically maintain its desired position in almost any sea. It was, at times, labeled a "lucky" rig,[64] and had just come from drilling the, then, world's deepest well. That well, in the Tiber Field, was 35,050 below the subsea surface, which, in turn, was 4,132 feet under water. This accomplishment even exceeded the specifications described by Transocean in its fleet status report as on April 13, 2010.[65] Longer than, and more than twice as wide as a football field, and able to operate in depths of up to 8,000 feet, it was universally regarded as up to the task of drilling at the Macondo Prospect.

Transocean had experienced some problems with the rig, including an evacuation in 2008 when a ballast system malfunctioned. However, it was still viewed as one of the world's top rigs. And, its crew and drilling master were very experienced, which was to be expected for a half-billion dollar asset that was contracted at almost half a million dollars per day.

Once such a drilling rig is in place, the process of drilling is not much different than a traditional land-based well. A section of drilling pipe is secured to a large and elaborate diamond-studded drill head. A machine rotates the drill pipe and the drill head while it circulates "drilling mud" down the pipe. The mud is used both to lubricate the drilling head and to carry away from the drill head the crushed rock debris as it proceeds.

Once the drill has penetrated rock equal to the length of one section of pipe, drilling is temporarily stopped, a crane system moves another pipe in place, and the new section is screwed on to the previous section using interlocking threads of pipe of a diameter than can range from two or three inches to a dozen inches or more. Drilling can then resume for another ten to forty feet until another section must be attached.

State-of-the-art drilling heads can now be steered and directed, allowing the operator to drill vertically, horizontally, diagonally, or in any other fashion not inconsistent with the lengths of pipe that must follow.

After the drill penetrates equivalent to a number of ten to thirty lengths of drill pipe, the drill pipe must be withdrawn and a length of casing inserted and cemented to the surrounding rock. With the casing in place, drilling can again commence with a slightly smaller drill head without the fear of the rock around the newly placed drill hole collapsing and jamming the drill head in place.

This process can be repeated as the drill head is steered until it reaches the reservoir. Drilling companies have successfully penetrated upwards of seven miles of rock, both on land and in water more than two miles deep using this technique.

However, mapping of the rock below, and the occasional requisite borehole and seismic sensing, at times, when the drill pipe is withdrawn, cannot give a perfect picture of what the drill pipe might encounter. There is always some uncertainty in what dangers the drillers may fall upon. Because the pipe is connected directly to the area in which workers are operating equipment that spin the drill head, the most significant danger is the tapping of high pressure natural gas or oil that could travel up the drill pipe and blow out into the drilling area of the rig.

A balancing act

The potential for such a blowout is minimized by the use of drilling mud to balance from above the pressure of gas and oil from below. The volume and density of such mud is adjusted to maintain the proper balance. This balance must be maintained even as the drill pipe is removed for wellbore measurement and inspection. If you recall, it was the failure to inject sufficient mud when a drill pipe was removed that caused the Ixtoc I well to blowout off the coast of Mexico and Texas in the Gulf of Mexico in 1979.

To address this danger, a blowout preventer (BOP) is installed on top of the first length of cemented pipe casing protruding through the sea floor. This massive device, that can stand more than fifty feet tall and weigh hundreds of thousands of pounds, is essentially a series of emergency valves that can pinch down the casing and drill pipe in the case of a blowout.

The cementing of the casing, an accurate mapping of the subsea floor, periodic centering of the pipe in the casing upon completion, the use of mud, and a heavy and viscous fluid that is chosen to be sufficiently heavy to counteract the pressure of the well, are each used to provide for multiple redundancies and protections to avoid a well experience condition that could lead to a blowout. While each of these redundancies work with each other to keep the well safe if it encounters a highly pressurized reservoir or pocket, the combination of mechanisms should still provide sufficient protection if one or another of the other protections fail. And, should everything fail, a blowout preventer with a dead-man's switch, which is automatically activated should communications with the rig itself be compromised, will pinch down the well pipe in three different places to ensure that even a localized pipe joint will not get in the way of successfully sealing off a runaway well.

A number of these systems must fail for a rig to enter into a dangerous blowout condition. Such a failure of multiple systems had never occurred on a deep-sea drilling rig – until April 20, 2010.

13
What Went Wrong – A Congressional Perspective

Like the fickle Macondo village of fictional fame that bounced between feast and famine, the Macondo Project would not easily give up its fortunes. With the reputation as a difficult well, only those drillers most confident could vie for an expected bounty worth perhaps $40 billion.

There are an incredible number of variables that must be managed when drilling a well. Before offshore wells, drillers were concerned over unknown and changing geology, striking a pressurized field that could cause a blowout, balancing the pressure of the field with the correct amount and density of mud, equipment breakage, especially at critical times, human error, and, of course, the possibility of explosion or highly flammable liquids catching fire.

As more accessible and shallower fields were depleted, drillers had to manage even greater risk by drilling longer at greater depth for more highly pressurized crude farther offshore. If an accident occurred, help could readily arrive to the wellhead on land-based wells. Offshore wells are much more vulnerable.

As the easy oil wells were exhausted, drilling proceeded in harsher climates. The massive Prudhoe Bay field on the North Slope of Alaska and abutting the Arctic Ocean, added the complication of temperatures one hundred degrees colder than was the norm for man and machine. These extreme variables increased both the risk and the expense of oil exploration and extraction. However, a growing population and steadily increasing price of oil provided the financial incentive to incur the greater risk and expense.

As global oil demand continued to rise, and low risk reservoirs were exhausted, oil companies used technology and accumulated expertise to successfully drill, first in shallow wells below the surface of the ocean, and, increasingly in wells deeper under the sea floor and in deeper water. As they did so, the number of variables rose. Pressures at an ocean depth greater than a mile are the equivalent of 200 atmospheres or more, or 200 times the ambient atmospheric pressure on earth. The pressure of a well drilled six miles below that subsea surface can be more than 1,000 atmospheres. And, while the temperature at the sea floor may be only a few degrees above freezing, the temperature rises rapidly as the drill pipe penetrates toward the reservoir.

The temperatures and pressures encountered at these depths are sufficient to contain the various gases associated with crude oil. However, if these gases come in contact with water at such depths and temperatures, icy hydrates can form, and freeze valves and equipment in place. Should this liquid combination of crude oil and liquefied gases move past frozen valves and meet the ambient atmosphere, large amounts of explosive natural gas can form spontaneously.

These new risks must be managed well beyond pressures that would afford direct human intervention, should anything go wrong. For the first time, wells are drilled, valves and blowout preventers are installed, repairs are made to equipment broken, and casings are inserted and cemented in conditions where only remotely operated vehicles (ROVs) can operate. And, if these new risks are not managed perfectly, repairs must be made, often ploddingly slow and carefully, far away from direct human aid.

This is the environment ExxonMobil found itself in just a few years before BP began drilling at the Macondo Prospect. Their Blackbeard West well in the Gulf of Mexico was almost six miles below the sea floor, and deeper than almost any other well in the world. ExxonMobil began to run into pressures that it believed it could not successfully manage. It abandoned the well just a couple of thousand feet short of the reservoir.

The well was soon successfully drilled by a competitor and will likely yield upwards of $150 billion in crude oil revenue as a reward. Clearly, ExxonMobil would have liked to have successfully managed the risk that could yield so much revenue for just a little incremental risk and expense. This is the tense, supercharged combination of human decision-making, some of the most sophisticated technologies in the world, and the harshest environments humans have ever attempted to master, on earth or in space, that has increasingly become the norm in our global pursuit for new oil.

BP, and its drilling partner Transocean, were determined to master this environment at the Macondo Prospect. Such newer, bigger, and deeper prospects are unproven every time, almost by definition. In their pursuit of oil, exploration companies increasingly have to operate beyond the scope of past practices, and have to manage risk never before managed. All the while, the various partners in this process must experience the financial pressures that measure expenses in upwards of a million dollars a day, and contingent rewards valued in the tens of billions of dollars.

Companies conducted these inherently risky procedures under the oversight of the Minerals Management Service (MMS), since renamed the Bureau of Ocean Energy Management, Regulation and Enforcement, a branch of the U.S. Department of the Interior that monitors and permits every facet of the drilling operation. The approval alone for such a drilling project requires the submission of a complete well plan that has been uniquely designed and engineered to best accommodate the depths, topographies, and challenges every deep water exploratory well presents.

A history with Transocean

While the Macondo Prospect was a challenging well, it was by no means at the deepest depth, nor the deepest reservoir, to be drilled by Transocean's Deepwater Horizon rig. When Hurricane Ida had knocked out Transocean's Marianas rig at the Macondo Prospect in November of 2009, there was little doubt that its replacement, the Deepwater Horizon, was up to the job.

Deepwater Horizon was considered one of Transocean's premier semi-submersible rigs. It was a fifth generation rig, and was relatively new. There are now sixth generation semisubmersible platforms. However, its dynamic stabilization system and computer system retrofits made it one of the Gulf of Mexico's premier platforms.

There had been a few safety violations, and one inadvertent release of oil, since its construction. However, none of these circumstances rose to the level that would significantly jeopardize safety or portend to the calamity that would soon engulf the rig. Indeed, the Deepwater Horizon was considered one of the Gulf of Mexico's safest rigs, and was even hosting a visit by executives to celebrate seven years without a lost-time accident the day the Deepwater Horizon blowout occurred.

Transocean is a company that specializes in deep water drilling. Their fleet is state-of-the-art, and they dominated the industry for deep water drilling, even before their absorption of their major competitor, GlobalSantaFe, in 2007. Transocean rigs were the rigs of choice for BP in the Gulf, where they employ four of Transocean's fourteen Gulf of Mexico rigs. BP chooses drilling rig contractors with better-than-industry-average safety records, and has consequently earned better-than-industry-average safety records for the last half dozen years.

Despite Transocean's involvement in the failures that culminated in the April 20th fire on the Deepwater Horizon rig, BP chose a Transocean rig to play a major role in the recovery and securing of the runaway Macondo Prospect well.

Transocean has not been without its problems. *The Wall Street Journal* reported that MMS data indicated problems with a Transocean blowout preventer in 2006, and with cement seals in 2005.[66] Indeed, the failure of cement seals, typically installed by specialist cementing subcontractors, caused more than half of the catastrophic well failures in deep water wells in the Gulf of Mexico.[67]

A problematic blowout preventer

A minor spill resulted from a failed blowout preventer in 2006.[68] On February 12 at 6:30 a.m., the Transocean rig Discoverer Enterprise, also commissioned by BP, detected unusual flows through their blowout preventer. As was the ill-fated BOP at the Deepwater Horizon well, the BOP was of a Cameron International design and was under a routine maintenance program of Transocean design and administration. Unfortunately, the BOP had not undergone the annual inspection mandated by Transocean. The incident occurred five months after an inspection was due, and the clogged valve resulted in a small oil discharge, classified as less than fifty barrels, into the Gulf of Mexico.

The failed blowout preventer was subsequently determined to have debris that had penetrated the blowout preventer valve mechanisms and did not allow them to operate properly. The MMS investigation determined that inadequate maintenance of the BOP was responsible for the incident. However, the incident was noted against BP since it was the responsible party that had employed Transocean under contract to operate the drilling rig.

A federal investigative panel recently challenged Transocean on its maintenance of the blowout preventer implicated in the Deepwater Horizon spill. Transocean subsea superintendent Billy Stringfellow had not heard about reported persistent leaks in the blowout preventer, but testified to the Coast Guard investigative panel that he thought the leaks were not significant enough to be reported anyway.[69]

Stringfellow also testified that the blowout preventer had not undergone its regular certification as required by federal regulations. Instead, Transocean chose to monitor the function of the BOP rather than periodically disassemble and inspect it. To allow Transocean to test the BOP in place, it had to modify the device so that certain valves designed to operate the emergency ram would instead control "test rams" that were unable to pinch the drill pipe as originally designed.

This modification was described in a 2006 article entitled "Subsea test valve in modified BOP cavity may help to minimize cost of required BOP testing"[70] in Drilling Contractor magazine and was cowritten by Transocean's Gary Leach as a way to save money for drilling contractors. After the misdirected plumbing was acknowledged, Stringfellow later confided that the replumbing of the BOP by Transocean meant the BOP was plumbed wrong. Consequently, the efforts by the remotely operated vehicles could not manually shut down the well in the days after the rig fire and collapse.

In addition, on March 23, 2011, the U.S. Government-appointed Deepwater Horizon Joint Investigation Team that released a forensic report on the damaged blowout preventer.[71] The report concluded that flaws in current BOP designs may have allowed the well pipe to buckle inside the device. Such buckling of pipe under high oil pressure may have prevented the BOP shears to function properly. This conclusion shifts some culpability from BP and onto Cameron International, the BOP manufacturer.

A challenging well

Every extreme deep water well presents its challenges. It is impossible to drill a well a mile under the surface of the ocean, and two or more miles through rock, sandstone, and sand to a reservoir of oil without facing unique challenges. Consequently, each exploratory well is designed uniquely, and permitted specially by the MMS. The exercise is one of managing the risks and unknowns as they present themselves, and adapting the plan, and seeking approvals for departures from the original plan, on an ongoing basis.

The degree of uncertainty is proportional to the increasing complexity and extremity of modern deep water wells. What may be routine and manageable challenges now would have been regarded as almost impossible obstacles just a decade or two ago. In these respects, the challenges of the Macondo Prospect were not the most extreme ever encountered. There have been wells that have penetrated more rock, and wellheads that are deeper underwater.

For instance, the fields off of Brazil are considered ultra-ultra deep. This classification is reserved for wells ranging from 10,000 feet of water depth and reservoirs 25,000 feet below the ocean surface. At these water depths, pressures well exceed 300 atmospheres, or about 5,000 pounds per square inch. Only an industry characterized by the world's most sophisticated, and hence complicated, technologies could function at such depths and pressures.

Brazil's new reservoirs, almost five miles below the ocean's surface, are also challenged by ocean-floor substrates that have, until recently, defied successful drilling. The Brazilian drilling platforms must penetrate sand, salt of various levels of porosity and density, and rock, and do so two hundred miles offshore, or about four times further off shore than the Macondo Prospect.

However, in combination, the Macondo Prospect was a challenging well, and presented its share of frustrations along the way. Indeed, BP engineer Brian Morel labeled it "a nightmare well which has everyone all over the place" in an internal BP email that was recovered by the U.S. House Committee on Energy and Commerce.[72] To be fair, this email was sent by a BP engineer to another BP employee as an apologetic rationale for last minute changes in design as a consequence of an earlier drill equipment failure.

A sequence of events

The Deepwater Horizon blowout, and the subsequent environmental remediation, has already become the most studied major private sector technological and environmental failure.

Other major technological failures have been well-studied before, most notably the investigation of two Space Shuttle in-flight catastrophic failures. Such major investigations that occurred in the aftermath of the destruction of the Space Shuttle Challenger 73 seconds after liftoff in 1986, or the Space Shuttle Columbia as it descended over Texas in 2003 on the way to its Florida landing site, were motivated by a desire to better understand the root of catastrophic failures in highly complex systems designed with many levels of redundancy.

The investigations into the Deepwater Horizon spill, first by the U.S. House Committee on Energy and Commerce, then by BP itself, and subsequently by the U.S. Coast Guard and by the U.S. Department of Justice, share the same motivation. Each asks "What sequence of failures occurred to create such a catastrophic spill when wells are designed to manage risk and prevent oil spills?" Each study also reinforces one prevailing theme. In well-designed systems, there is not one single factor that "caused" the unexpected outcome. Rather, each failure reduces the levels of redundancy by one so that subsequent failures become more problematic.

The accident in a nutshell

The Macondo Prospect well was challenging, as are all wells in extreme conditions. The job of BP Exploration, the designers of the well, and the drilling company Transocean that would use their drilling rig and staff to drill the exploratory well, was to tap, and then cap, the Macondo Prospect. On the day of the explosion, the exploratory well had been tapped, and was being prepared for a cement seal that would act as a cap until another group would return to connect the well to a network of oil pipes and manifolds that crisscross the seafloor of the Gulf of Mexico.

This well, in particular, had ongoing challenges with escaping hydrocarbons, typically methane gas. In the evening of April 20, 2010, at 9:45 p.m. local (Central standard) time, as the drilling team prepared to cap the well and were packing up in preparation of departing the site, gas under high pressure shot through the drill column all the way to the drilling platform at the surface. The resulting explosion, fire, or attempts to escape the platform towering a hundred feet above the ocean surface, caused the death of eleven platform workers. Their bodies were never found.

The last few moments

The President's National Commission on the BP Deepwater Horizon spill and Offshore Drilling provided a gripping account of the last moments on the rig. The following describes the observations from the report of the state of mind of the rig crew in the day of the explosion.[73]

Early in the morning of the day of the blowout, the cementing job designed to finish the bottom of the well, at the reservoir, was completed. The cementing engineer reported to his boss that the job had gone well. However, the night before, there had been calls between the cement crew and their head office about some concerns of natural gas penetration.

The BP drill plan had a contingency in case the cement crew reported the job did not proceed to specifications. BP could perform a cement log to map the effectiveness of the cement job. Alternately, the MMS also allowed pressure tests to be performed in lieu of a cement log. Indeed, BP flew out a cement log team in case a cement log was needed, but flew the team back after conferring with its cement contractors. Instead, BP would perform the two pressure tests.

Clearly, if BP had known or better understood the pattern of problems with Halliburton's deep-sea cementing specifications, or if there was better communications between BP and Halliburton in the last days of centralizer and cement design, a cement log would have been most prudent. Instead, BP employed alternative tests.

The first test, a positive pressure test, imposes positive pressure on the steel casing and seals to ensure they hold and do not bleed down pressure at an unacceptable rate. The positive pressure test was successful. This milestone gave the contractors and crew a sense of comfort that the job had gone well, the well was sealed, and the process of moving to the next job could begin.

After BP performed a positive pressure test of the well to ensure it could hold well pressures without a leak, the BP company man, Robert Kazula, was dissatisfied with the team's interpretation of the results, and asked for additional testing. Transocean rig leader Jim Harrell demanded an additional negative pressure test.

Meanwhile, two executives from Transocean's Houston office, and two BP executives from Houston arrived by helicopter for a "management visibility tour."[74] As the senior executives were touring the facilities, the Transocean team began the negative pressure test.

The test did not go perfectly well. The test requires the crew to reduce pressure to the well and see if that reduced pressure can be maintained. If so, the team would have confidence that no gases, mud, or hydrocarbons were bleeding into the drill pipe. However, the negative pressure test showed a slow but consistent buildup of pressure.

By the completion of this additional test, the Transocean subsea supervisor, Chris Pleasant, discovered that up to 60 barrels of mud could have been lost during the pressure test. This signaled to Pleasant that there may be a problem with well integrity, a concern shared by Transocean tool pusher Wyman Wheeler.

By this time, though, the crew's shift had ended and Wyman Wheeler passed duties off to the next crew. After discussion with the relief Transocean tool pusher, Jason Anderson, Don Vidrine, a BP representative, asked the Transocean drillers to perform an additional negative pressure test. This test ended a little more than an hour before the first signs of imminent trouble. Tragically, the results of these tests were fatefully misinterpreted. Readings unexpected by the BP representative were reportedly erroneously explained away by Transocean tool pusher Anderson as due to a "bladder effect."[75] Upon completion of the positive and negative pressure tests, BP well site leader Don Vidrine instructed his Transocean counterpart Kaluza to begin to displace drill mud in the well pipe with seawater in preparation for well closure and abandonment.

Meanwhile, Transocean rig Captain Curt Kuchta was demonstrating the rig function to a few high level executives that were touring the rig. It was just past 9:00 p.m., and the VIP crew was in the bridge receiving a description of the advanced technologies that hold the rig fixed above the wellhead through the use of six high-powered thrusters. The visitors were even given a turn on the practice simulator to see if they could, theoretically, keep the rig in place under 70 knot winds, 30 foot seas, and a third of the thrusters offline. Positioning tonight, though, was a simple matter. The seas were glassy calm and the night sky was filled with stars. Instead, the discussion was over BPs concerns that there was a backlog of 390 safety items that would require almost two man-years to remedy, based on its recent safety audit of the Transocean rig.[76]

While Miles Ezell, Transocean's senior tool pusher, remained concerned, despite the reassuring "bladder effect" theory put worth by relief tool pusher Anderson, Ezell retired to his cabin at 9:30 p.m. Minutes later, the first kick of gas began to rise toward the rig. Transocean's Pleasant and Harrell were completing paperwork and wrapping up their duties when, at 9:50 p.m., Ezell was called. The assistant driller Steve Curtis reported that "we have a situation" and that the well was blowing out.[77] Mud was beginning to spew out of the crown of the drill pipe, and Ezell's relief tool pusher Anderson needed his help to shut in the well. Moments later, an intense gas kickback and explosion killed Anderson before Ezell could leave his cabin.

About the same time, a rig worker asked Transocean supervisor Pleasant about seawater spewing onto the deck of the rig. For a moment, Pleasant failed to grasp the gravity of the situation, and instead remained fixated on his computer screen. When a minute later the worker said he also sees mud rising, Pleasant called the rig floor but was unable to reach any workers. He immediately ran to the rig floor just as Captain Kuchta first saw fluid pouring onto the rig. At the same time, survivors reportedly heard a loud hiss of gas just before the first explosion rocked the rig. A fire and second explosion soon followed.

Chris Pleasant, the Transocean subsea supervisor and the supervisor in charge of the blowout preventer, later testified that he implored the captain to immediately hit the emergency disconnect switch to separate the rig from the riser. The captain refused his request. Pleasant himself would trip the switch thirty seconds later. However, by then, the switch no longer functioned.[78]

The rig fire burned uncontrollably for two days, despite efforts by water cannons from rescue ships to douse the flames through the spraying of millions of gallons of water. Thirty-six hours after the first explosion, the platform listed and sank. As it descended to the seafloor, it also took down the riser pipe that had remained attached on one end to the platform and at the other end to the blowout preventer on the seafloor. The partial or complete shearing of the riser pipe, at numerous spots, created the breach that resulted in uncontained oil that would spew from a chaotic web of 5,000 feet of 16" pipe that was by then zigzagging the bottom of the Gulf of Mexico.

A Congressional investigation

The U.S. House Committee on Energy and Commerce began investigating the oil spill almost immediately after the blowout occurred. On June 14, 2010, they summarized their investigation to that date in a letter seeking responses from Tony Hayward, Chief Executive Officer of BP PLC. In that letter, they described five circumstances that they believe may have contributed to the spill:[79]

a) Well design
b) Centralizers
c) Cement bond log
d) Mud circulation
e) Lockdown sleeve

As each of these issues is treated in turn, it must be noted that this hearing was organized around a prevailing premise. The committee postulated in its preamble address to BP CEO Hayward that BP engaged in risky practices because of a concern over cost overruns at the well. The committee noted in the opening of its letter to Hayward that "it appears that BP repeatedly chose risky procedures in order to reduce costs and save time and made minimal efforts to contain the added risk." The Congressional Committee postulated the following five potential causes for the accident.

Well design

With regard to the events leading up to the blowout on April 20, the committee postulated that BP chose a well design that was less likely to contain pressurized explosive gases that could escape into the well. This cavity, or annulus, between the drilled hole and the steel casing lining the hole, could allow oil and gas to penetrate the casing itself and find its way to the ocean surface.

As BP made final preparations to complete and close the well on April 19, the day before the blowout and explosion, it had decided to install one continuous casing, from the wellhead to the reservoir, rather than install an extension of the final steel casing it had placed earlier and which penetrated to within about eleven hundred feet of the reservoir. The Committee noted that, if BP instead "hung" an additional length of casing onto the last casing in place, the "tieback" joint between the two pieces of casing would have been cemented in place and would have acted as a redundant barrier that could have prevented hydrocarbons from travelling up the outside of the casing and penetrating into the interior of the casing. This redundancy would have been in addition to the cementing of the casing or liner at the bottom of the casing pipe where it met the oil reservoir.

To understand the significance of BP's decision, a fuller description of wellbore drilling strategies and cementing is necessary.

The bottom cement is a critical first line of defense in preventing hydrocarbons from blowing out in an uncontained manner. When a deep water well is drilled, it begins with a wide diameter of thirty-six inches or more. Each section might be a thousand or more feet deep and must be drilled a little larger than the piece of steel casing that will be placed down into the hole. Each such hole is rough, and may be through rock or sand that is loose and prone to collapse. The inserted steel casing of a smaller diameter than the hole itself is then bonded securely to the surrounding rock and sand using specially designed cement that is mixed to withstand the pressures at depth and bond well to the substrate material. This bonding of the casing to the surrounding rock gives the well integrity and blocks hydrocarbons held under pressure below from migrating up this annulus between the steel and rock.

The use of specially designed cement that would successfully bond sometimes loose rock to steel is critical. However, regulators know that most blowouts occur because of failures in cementing. Consequently, multiple layers of casing and cement are used. Once a section of casing is cemented into place, a slightly smaller drill bit that can fit into the now narrower steel-cased hole digs deeper and allows the drilling contractor to insert another piece of casing further down. Again, cement is used to stabilize this new casing in its hole, and to join the new casing to the casing previously cemented into place. Each of these casing joints provides another opportunity for pressurized hydrocarbons to enter the wellbore from the rocky sides. And, each cement job provides one more barrier to prevent such penetration from occurring.

As this process continues, the wellbore looks like an inverted wedding cake, with a wide casing at the top, and progressively smaller casings at increased depth. The Macondo well started with a hole in excess of thirty six inches and had tapered to less than ten inches at the bottom, almost thirteen thousand feet below a wellhead that was 5,321 feet deep.

Once this tapered design neared the bottom of the reservoir, BP had originally planned to drill, hang, and cement a last casing of 11-3/4" diameter before running a continuous production casing from the top of the well to the bottom. This approach would allow one more cemented link between the annulus outside the casing and the joint between the last two casings, as another barrier to prevent the migration of hydrocarbons from outside of the well to the inside of the casing.

However, BP applied to the MMS to deviate from its original well design. As it tried to put its final sixteen inch casing in place, it encountered resistance and could not place the casing deeper than 11,585' below sea level, 915' shallower than the BP design intended. On March 8, 2010, one day after it began to drill a sixteen inch hole and set its second to last liner in place, it again experienced greater-than-anticipated pressure, this time from the entrance of hydrocarbons into the drilling area. As it attempted to adjust for this pressure, its drill bit became stuck and it had to reroute its well and redesign its plan. BP would have to abandon its plan to place a 13-5/8" liner to a depth of 15,300' and instead was able to place the casing only to a depth of 13,145'.

This obstacle to their original design required them to seek approval for a modification of the well plan below 13,145'. MMS granted BP the deviation, and BP was forced to begin to use smaller diameter pipe below 13,145'. By 17,168', with more than a thousand feet of well drilling to go to reach the reservoir, it had placed its last casing, with a diameter of 9-7/8", or significantly narrower than the original 11-3/4" design it had initially submitted for approval. These changes were necessary to accommodate the well deviation because of the stuck pipe and the higher-than-expected pressures the drillers were encountering.

The drilling and placement of this last, narrower-than-planned, casing was completed, and drilling resumed to the reservoir region on April 2. Well drilling was completed on April 9. BP and Transocean then conducted tests of the surrounding substrate for five days.

Given the change in design of the lower well, BP had to redesign the production pipe that would be inserted into the well liner and would eventually be used to extract oil from the reservoir. Its original design called for a two part pipe – one liner that would run up to the last casing and would be cemented to the casing, and another "tieback" that would run to the wellhead. This process would take another three days to cement and union, but would provide one more barrier that would block pressurized hydrocarbons from penetrating the casings from outside, and would prevent an open annulus from running unobstructed over the extent of the well between the production casing and the liners.[80]

However, in the redesign, BP elected to substitute the original 9-7/8" pipe, with the tieback, with a 9-7/8" section at the top, and a reduced diameter 7" section below about 12,000', and no tieback redundancy. This solution would be quicker. The House Committee asserted that economics trumped safety in BP's revised and MMS-approved decision.

Centralizers

On April 14, BP received the results of a consultation request it made to Halliburton, its cementing contractor, to run computer simulations to determine if this new design would meet the higher pressures it had encountered and still provide sufficient zonal isolation to prevent high pressure hydrocarbon infiltration. The Halliburton 9-7/8 in. × 7 in. Production Casing Design Report concluded on April 14 that this redesigned solution would work, and specified 10 centralizers – six of a diameter of 9.875", and four with an eight inch diameter.[81]

The issue of centralizers is important because a production casing that is not held in the center of the drilled well for cementing will be difficult to clean and flush before cementing. If the casing is trapped up against the wall of rock, it is possible that the drilling mud used to flush the hole of loose rock and other material will not flush away all remaining loose debris before cementing. These remnants could remain adhered to the rock wall and may not be displaced as cement is injected into the cavity. This mud could then be pushed away subsequently by high pressure oil and gas that attempts to find a pathway through the annulus between the well and production casing and make their way to the wellhead. This migration of oil and gas around mud and cement is called channeling.

Actually, Halliburton, the cement contractor, ran a number of scenarios through the OptiCem program over a few days. One of the subsequent runs of the OptiCem program on April 15, based on updated survey data, respecified that 21 centralizers should be used to keep the longer production casing centered for cementing. An order for 15 additional centralizers was made, and the spacers arrived on the platform on April 16.

As new wellbore survey data became available, between April 14 and April 18, a number of centralizer and cementing recommendations were made by Halliburton. The report on April 14 concluded that the new well design with the one-piece production casing would work. Two subsequent analyses on April 15 demonstrated there would not be channeling of mud and gasses if 21 centralizers were employed, but that potential channeling would occur if 10 centralizers were used.

The House Committee documented a series of e-mails that went back and forth between well designers and managers at BP and the cement contractors at Halliburton.[82] Brian Morel, the BP engineer, postulated that the wellbore ran reasonably straight, and thus would need fewer centralizers. Another BP drilling engineer, Brett Cocales, pointed out that even a pipe under tension from the force of gravity will not necessarily lie perfectly straight, even in a vertical hole. However, he also agreed that if the decision had been made to go with six centralizers, it would probably work out anyway, so long as there was a good cementing job.

Cocales' comment that six centralizers would probably be adequate was based on an informal risk–reward calculation.[83] While such a calculation appears in error, in retrospect, Cocales was confining his comments to the decision already made, under the assumption that all other design safeguards remained intact. Of course, in retrospect, this calculation seems fatally flawed.

At the time of that analysis, rig management had expressed concerns over the type of centralizers, known as bow spring centralizers, delivered to the platform, given the possibility that slip-on stops attached to the centralizers may present problems the team had encountered on a previous job. The team on the rig decided to proceed without the additional bow spring centralizers.

After the production casing was inserted into the wellbore, Halliburton, on April 18, went on to perform a subsequent analysis that demonstrated seven centralizers would provide an adequate cementing job if a more elaborate, nitrogen-saturated cement were used. In that correspondence, Halliburton concluded that the April 18 design, with a reduced number of centralizers, was "our recommended procedure for cementing."[84] Earlier that day, the team had already begun inserting the production casing into the wellbore.

This final centralizer-cementing analysis determined that, with seven centralizers, a severe channeling problem would occur if higher than measured reservoir pressures were encountered. BP's own investigation later concluded that the drill team should not have proceeded to insert the production casing with a reduced number of centralizers until the contradictory and incomplete OptiCem analyses could be resolved. The investigation concluded that the well team should also have correctly identified that the additional 15 centralizers met specifications. However, the investigation team also believed that an insufficient number of centralizers, and the concomitant greater probability of channeling, was likely not the cause of a hypothesized cement failure.

In addition, the *New York Times* reported that Ronald Crook, a cementing consultant and chemical engineer who has worked for Halliburton, argued in 2008 that gas flow problems can be remedied by changes in the cement slurry and by a calculated placement of the cement in the annulus between the wellbore and the production casing. His article stated: "In wells with severe levels of gas migration, the risk of a gas flow problem can be reduced to a safe level by adjusting those other factors." It was possible that subsequent runs of OptiCem would have been able to optimize the cementing design, had BP and Halliburton communicated more effectively.[85]

Halliburton was subsequently implicated by a Presidential Commission charged with investigating the BP Deepwater Horizon spill. In a letter from an investigator for the National Commission on the Deepwater Horizon Oil Spill and Offshore Drilling, Fred Bartlit reported that Halliburton shared with BP that its original cement formulation was defective. However, after many redesigns of the cement job and mixture, subsequent test results had not been shared by Halliburton with BP. Halliburton nonetheless used the failed cement specifications in the critical Deepwater Horizon cement job that subsequently failed. The investigative committee reported:[86]

> We have known for some time that the cement used to secure the production casing and isolate the hydrocarbon zone at the bottom of the Macondo well must have failed in some manner. That cement should have prevented hydrocarbons from entering the well. For a variety of technical reasons that we will explain at the upcoming hearing, BP cemented the well with a nitrogen foam cement recommended and supplied by Halliburton. Halliburton generated the nitrogen foam cement by injecting high pressure nitrogen into a base cement slurry as it pumped that slurry into the well...
>
> Halliburton provided data from one of the two February tests to BP in an email dated March 8, 2010. The data appeared in a technical report along with other information. There is no indication that Halliburton highlighted to BP the significance of the foam stability data or that BP personnel raised any questions about it. There is no indication that Halliburton provided the data from the other February test to BP.
>
> Halliburton conducted two additional foam stability tests in April, this time using the actual recipe and design poured at the Macondo well. We believe that its personnel conducted the first of these two tests on or about April 13, seven days before the blowout. Lab personnel used slightly different lab protocols than they had used in February. Although there are some indications that lab personnel may have conducted this test improperly, it once again indicated that the foam slurry design was unstable. The results of this test were

reported internally within Halliburton by at least April 17, though it appears that Halliburton never provided the data to BP.

Cement bond log

The House Committee hearing on the response of BP did not challenge BP's role in a cementing job that was the responsibility of Halliburton, the cementing contractor. However, the House Committee did challenge BP's decisions to forego a cement bond log to fully test and document the cementing job.

As discussed earlier, regulatory agencies have stated that the majority of blowouts are caused by cementing failures. A cement bond log is a day-long procedure that uses imaging techniques to measure the integrity of the cement bonds of borehole to casing and casing to casing. BP had flown a team from Schlumberger, a major oil field service company, to the platform to perform the log, but decided to forego it and had the team return to the mainland on the morning of April 20, the day of the blowout.

The well plan does not require a cement bond log. Rather, MMS specified that the oil company must:

(1) Pressure test the casing shoe,
(2) Run a temperature survey,
(3) Run a cement bond log, or
(4) Use a combination of these techniques.[87]

BP had instead elected to perform a pressure test. The House Committee speculated that BP may have balked at the incremental $118,000 cost, and 9–12 hours if it proceeded with the cement log test.

MMS stipulates that the top of the cementing should extend five hundred feet above the highest identifiable hydrocarbon zone. This distance is a compromise that allows sufficient distance to ensure proper cement penetration without running the risk of cementing so high that the next casing string is sealed. If it were sealed, pressure build up during production could cause the casing to burst.

BP's own best practice for isolation of zones in the wellbore calls for cementing to one thousand feet above any permeable zones of substrate, with centralization extending another one hundred feet farther. If these conditions are not met, the BP best practices states that cement penetration should be measured using a cement evaluation log.

While the team did meet the requirements of MMS and federal law, BP investigators subsequently concluded that the team should not have departed from published BP best practices without conducting a formal risk assessment. The investigative team recommended that BP and its cement contractor communicate more effectively and in a timelier manner so that both parties can accurately assess and balance cementing risks.

Instead, BP met MMS and federal standards by performing two types of pressure tests. However, the drill operators and BP fatefully failed to properly interpret one of these tests, as will be more fully described in the timeline to the blowout later.

Mud circulation

The oil industry uses the term "mud" to refer to a material that serves many purposes. A mix of synthetic hydrocarbons and granulized rock, mud can lubricate the drill head and flush ground rock at the drill point toward the wellhead and, subsequently, to the platform or mud ship. Mud can be mixed to a desired density to balance the pressure of hydrocarbons from below. Because mud is heavier than hydrocarbons, a specified column height of mud from above can balance the pressure of oil from below. Finally, mud can also be used to flush materials from the well liner and the annulus between liner and the rock and sand wellbore or the production casing in anticipation of cementing.

A full cleaning of a well casing, with mud forced down the production pipe, and up the space between the pipe and the liner, is called a "bottoms-up" mud circulation. In addition to cleansing the bore of debris before cementing the production casing in place, it also moves to the platform any gas pockets, and will indicate to the mud crew whether additional hydrocarbons are infiltrating the well. Mud monitoring, and the bleeding of gases away from the platform, are essential specialties in the drilling process.

In order to ensure the mud flows well for such a cleansing, the American Petroleum Institute recommends circulating a volume of mud equal to 1.5 times the volume of the space between the lining and the casing, whichever is greater, for a bottoms-up cleansing.

The House Committee quoted BP's operation plan as recommending "a bottoms-up mud circulation of one casing and drill pipe capacity, if hole conditions allow."[88] The House speculated, based on testimony from a Halliburton account representative, Jesse Gagliano, that the BP well managers instead chose to circulate 261 barrels of mud. The approximate volume of the production casing alone is 1264 barrels, or approximately five times the circulated volume of mud. It was speculated that this reduced circulation would have saved an additional twelve hours of platform labor.

The possibility that a partial mud circulation rather than a complete bottoms-up contributed to incomplete cement bonding or gas channeling remains an open issue. The fragility of rock and sand material at the bottom of a well that must be sealed may have dictated a cleansing that circulated adequate mud in only the region of the cementing. Such a partial cleansing with mud that was not subsequently brought to the platform for inspection of debris or gas infiltration made it difficult to detect any conditions that may lead to bonding failure or gas infiltration. These issues will require further investigation.

Lockdown sleeve

Finally, the House concluded that BP may have left off a lockdown sleeve that helps secure a large gasket at the top of the wellhead and prevents movement upward of the production casing as a consequence of unexpected reservoir pressure or thermal expansion of the production casing as hot oil moves through it or as it becomes buoyant under certain conditions. BP's drilling contractor, Transocean, implicated BP for this omission.

However on the day of the fateful explosion, BP had been seeking permission from MMS to install its cement plug that would secure the well until production at a level deeper than previously approved. MMS regulations require 300 feet of cement to be set in the pipe as a plug, should the shoe track fail. BP felt it prudent to place this plug at 3,300 feet below the ocean floor, deeper than specified by the MMS.

If permission had been granted for the revised cement plug location, BP had planned to flush any remaining drilling mud from the riser pipe above the wellhead with seawater, install a cement plug, and then install the lockdown sleeve. If MMS did not grant permission, BP would then install the lockdown sleeve and install the cement plug at a shallower point.

Because the well securing process was not yet complete when the fatal Deepwater Horizon blowout and explosion occurred, it is unclear if an uninstalled lockdown sleeve was a fateful omission. Subsequent analysis by BP has established that production casing lift was not a problem. BP has determined that there may have been other issues with the seal at the top of the wellhead that could have occurred at an earlier stage in drilling.

14
Lessons for BP from More Considered Reviews

There is little doubt now that there was not one single engineering, equipment, or human failure that led to the Macondo disaster. The oil industry now well knows that complex systems create the possibility of complex failures. The Challenger space shuttle disaster has taught engineers that risk management must expand in proportion to the complexity of engineered systems. What is less understood, even today, is the role of management systems that can rival in sophistication the engineering systems designed to mitigate risk. After all, as BP's own engineering analysis shows, even multiple engineering safeguards and sophisticated data acquisition may fail if humans cannot properly manage an enterprise growing in complexity.

An engineering analysis

BP subsequently conducted an internal investigation of the failure in the hope that it would improve best practices at its other deepwater wells, inform the industry, and offer a resource to Congressional and subsequent Coast Guard and Department of Justice investigations.

The BP investigation team noted that its report should not be viewed as an effort to assign responsibility or shed liability. Instead, under the leadership of Mark Bly, BP's Group Head of Safety and Operations, a team was assembled to be independent of BP's other emergency response teams, and to draw upon the public record, BP documents, correspondences with its contractors Transocean, Halliburton, and Cameron International, among others, and 50 internal and external specialists. Its goal was to employ its expertise in safety operations, deep-sea drilling, well control, cementing, wellbore modeling, blowout preventer operations, and risk analysis to better understand what lead to, and what could have prevented, the blowout and explosion on April 22, 2010.[89]

The investigation shared with, and departed from, some of the conclusions of the House Energy and Commerce Committee. For instance, the BP investigation team concluded that the well redesign after the pressure problems and wellbore diversion would not have created a substantial increase in well risk. And, while the investigation found fault in the number of centralizers employed to center the production casing in preparation for cementing, this issue should have been substantially mitigated through an optimized cement design.

Indeed, this practice of a reduced number of centralizers is not uncommon in Gulf of Mexico deepwater wells, as the Halliburton executive, Jesse Gagliano, recently testified. His belief that the reduced number of centralizers may not have been contributory to the blowout was recently reinforced by Halliburton vice president, Thomas Roth. Commenting on the cementing based on BP's revised design, he noted to a National Academy of Engineering hearing that: "We didn't see it to be an unsafe operation as it was being executed." BP drilling engineer, Kent Corser, went on to note that oil and gas migration past centralizers was not the cause of the blowout. Investigation of the well demonstrated that oil and gas migrated up through the drill pipe, not up the annulus along the side of the well.[90]

The BP investigation also concluded that the drilling team should have performed a proper analysis of the consequences of foregoing a cement log. While a well pressure test met federal statutory requirements as administered by the MMS, it departed from published BP best practices. Consequently, a full risk assessment should have been performed so that any risk mitigation options could have been explored including, perhaps, a cement log.

The investigation team did not address whether a more complete mud circulation would have improved the cementing operation or indicated growing gas infiltration problems. Finally, the BP team noted that the placement of a lockdown sleeve was anticipated in the sequence following the cement plugging of the well. The blowout and explosion occurred before this step could be initiated in its predefined sequence.

The investigative team concluded that none of these factors were critical in the chain of failures that caused the uncontrolled blowout and spill. Rather, the BP investigative team identified eight key barriers that were breached. Each played a critical part in the blowout and explosion. If any one of these eight barriers performed as designed, the disaster would have been averted.

Eight breached barriers

1) The bottom cement barrier did not prevent hydrocarbons from infiltrating the space between the well lining and the production casing.

The House hearing focused on BP's decision to not perform a well log. However, such techniques to measure cement penetration may not have successfully detected channeling of gases through the annulus. There is no substitute for a properly performed cement job, even if subsequent testing may or may not detect cementing imperfections. As reported, the majority of well blowouts occur because of cementing failures. While subsequent cement remediation or additional zonal isolation may reduce or eliminate the problems associated with an inadequate cement job, it is clear that successful cementing is critical to long-term well viability.

The sophistication of cementing has increased dramatically over the last two decades. As wells go deeper underground or below the surface of the ocean, the pressures that the cement must balance when injected, and overcome when hardened, become larger. Cement formulae and techniques must be designed to take into account many associated variables.

The primary force that the cement must first balance, and then overbalance, is the pore pressure from the drilled rock, salt, or sand. A pore pressure is the pressure the rock, and the water and hydrocarbons it contains, exerts on column of mud that keeps collapse of a drill hole at bay. An overpressured well, such as the Macondo Prospect, has a pore pressure that exceeds the weight of seawater from above. Consequently, the correct density and height of mud must be used to balance this high pore pressure.

However, the pore pressure is not constant, and may not vary evenly with column height in the same way as does the balancing mud. This fracture pressure gradient must also be taken into account as a cement mix is designed to displace the mud, balance the pore pressure gradient, and, as it hardens, provide a permanent overbalance that can keep the hydrocarbons from entering the well at alternate locations.

The density and pressure of the cement slurry injected into the bottom of the well is critical to ensure cement penetrates the proper gap between the wellbore and the casing. The volume of cement injected should displace an amount of mud that can then be measured as the displaced mud flows through the annulus and up to the surface. A proper balance between the cement injected and the mud displaced will indicate to the cementing contractor that the cement was properly injected, assuming that failures of the cement in filling desired regions was not made up by cement flowing in unanticipated regions of the well. The density of the cement is a critical factor in ensuring cement permeates its intended region. All of these factors must be balanced and optimized for a cement site that cannot be seen and which lies miles beneath the surface of the earth.

In fact, there is not simply one cement that is injected in such a well. Cap cement, tail cement, spacers, and foam cement are all used at different stages in the cementing process to ensure a proper seal. In the Macondo Prospect well, the engineers at Halliburton and the BP well team were preoccupied with ensuring the correct balance of these various factors.

This cement density was also critical because the Deepwater Horizon drillers had previously experienced unexpected loss of mud when drilling the bottom of the wellbore. A calculation called the Equivalent Circulating Density (ECD), measured in pounds per gallon, had to be reduced because mud was migrating either into the reservoir or fractures in the surrounding rock. A reduction in mud density managed to reestablish circulation of mud upward, rather than loss of mud to the well.

The BP investigative team concluded that the challenges of cementing, given the depth, pore pressures, fracture gradient, and well wall materials, preoccupied the cementing team. As a consequence, other critical aspects, such as additives to prevent hydrocarbons from mixing with the cement, and the mix of the cement slurry may not have received adequate attention.

In addition, the investigators believe that cement may have been lost during the final stages of injection. The Halliburton protocol called for a higher cement density in the latter stage of cementing than the well had been able to support in earlier mud injections. The volume of cement injected, at 61 barrels, was relatively modest. The cementing analysis showed that only about 3 barrels of mud were lost during cementing. Small losses of cement may be indicative of a successful cement job. However, there could also be other factors that can artificially create small losses. It is difficult to know precisely what is occurring at a depth more than three miles out of the reach of direct human observation. Consequently cement failures remain the single leading cause of well blowouts.

Also, if significant cement was lost near the end of the run as the cement density was increased, the cement would have little impact on the ultimate goal of isolating the bore from the surrounding sands. The investigators concluded this factor did not contribute to the accident.

However, the investigators were concerned with the mix of the cement slurry. The primary cement used at these depths is actually a foam that includes nitrogen gas under high pressure to allow the cement to flow better into the cavities and pores it must fill. The OptiCem program used to design the cement job called for a slurry that held 18% to 19% nitrogen by volume at the depth of the cement job. Because nitrogen is compressed at the high pressures encountered at the bottom of the well, this specification called for a 55% to 60% nitrogen mixture at 1,000 pounds per square inch when the cement is mixed at the surface.

An independent lab test by CSI Technologies, as commissioned by the BP investigative team, determined that a stable cement mix with the specified concentration of nitrogen could not be attained. The test results concluded that the cement used would have been unstable and would have resulted in nitrogen breakdown.

As discussed previously, a commission appointed by the U.S. Administration to investigate the causes of the spill noted that Halliburton discovered through tests that its cement specification had failed, but had not shared all such test results with BP. Halliburton was especially implicated in not sharing the results of later failed tests after the well plan and cement job had been respecified.

If the subsequent tests by CSI and Chevron prove to be affirmed, the cement instability would have weakened its ability to isolate the casing from hydrocarbon infiltration and may have also weakened another cement barrier in the shoe track, to be discussed next.

2) Fail-safe barriers in the shoe track did not isolate hydrocarbons.
At the bottom of the production casing is a device called a shoe track. Its purpose is to channel hydrocarbons from the reservoir through the production casing to the wellhead. As the joint BP-Government science team performed the static kill of the well on August 4, 2010, they concluded that pressurized hydrocarbon made its way up the production casing, not through the annular gap, called the "annular-A", between the casing and the steel well liner. Because the production casing is contiguous from the bottom of the well to the wellhead, hydrocarbons must have made their way through the cementing job and into the shoe track. However, the shoe track is designed to act as another barrier to prevent further hydrocarbon infiltration. The shoe track must also have failed.

The shoe track, or float joint, is an insert in the casing at the bottom tip of the production casing. The shoe track is sealed by the bottom cement job that is designed to bond the outside of the casing to the surrounding rock and sand formation. During the final cementing job, the shoe track guides the cement inside the bottom of the production casing that feeds the outside cement in the cementing process. By maintaining this internal reservoir of cement, the cement team can be more confident that a miscalculation has not forced the outside cement too far into the surrounding cavity that it has partially evacuated the lowest portion of the cement job. This extra reservoir of cement helps ensure a complete outside cement job.

This shoe track also uses a pair of one-way check valves that allow cement and mud to move down the track to fill the voids, but prevent cement, mud, or, in the case of a cement job failure, hydrocarbons, from coming back up. These flapper valves are industrial versions of the one-way flapper valve you can find in the water reservoir of a bathroom toilet. In theory, it prevents the flow of any material back up a well until the well is ready for production.

For hydrocarbons that may have breached the outside cement job to make their way up the production casing, both the shoe track cement job and the series of two independently operated flapper valves must have failed.

The cement in the shoe track could have failed if it had been mixed with foreign materials, such as mud or hydrocarbons, as it tried to set; it was mixed with nitrogen that may have migrated out of an unstable outside cement job; improper design of the shoe track cement; or some combination of these factors. Certainly, Halliburton's reported failure to test the cement that appeared to have failed may be a fatal flaw that precipitated the worst offshore blowout in history.

However, even if the shoe track failed, either of the two flapper valves should have stopped hydrocarbons from migrating any further. While the shoe track is used to pass mud through the well and balance the hydrocarbons from below, these valves are held open by a steel "autofill tube." This tube allows surges to move mud up the casing and ensures a safer balancing of forces in the well. Once the well is ready for sealing, the tube is pushed out, which allows the flapper valves to operate normally.

This autofill tube is activated by forcing mud through it with a differential pressure of between 400 psi and 700 psi. This additional pressure causes mud to flow through the autofill tube in a way that creates a pressure that forces the autofill tube down and out of the way of the flapper valves.

When the team attempted this conversion to push the autofill tube out of the valves, they found they needed a much higher pressure to establish flow. When, after nine attempts, they were able to establish flow, the flow did not exceed 4.3 barrels per minute, which is under the 5 to 7 barrels per minute that Weatherford, the valve manufacturer, calculated would be necessary to move the autofill tube. It is possible that the increased pressure merely unplugged a clog in the bottom of the shoe, and did not successfully displace the autofill tube. If not, the flapper valves would not function as designed to prevent subsequent hydrocarbon flow in the event of outside, and then inside, cement failure.

The investigative team also noted that the valves could have been damaged in a subsequent incident in preparation for cementing. Above the valves is a seal called a wiper plug that is designed to separate cement from the fluids that are used to flush the area in preparation for the cement job. This seal is designed to rupture between 900 and 1,100 psi, but did not burst until pressure of 2,900 psi was applied. The investigators speculated that this surge in pressure may have damaged the flapper valve. Further tests are necessary to confirm this hypothesis.

Regardless, this second barrier, if functioning as designed, should have prevented a failure of the first barrier from causing an upward flow of oil and gas.

3) Results from the negative pressure test were improperly interpreted.

The two final stages in securing a well for later production involve cementing the bottom of the well to ensure the well lining is securely sealed in the surrounding rock, and cementing of the top of the well to secure the well until production will commence. Cementing of the bottom of the wellhead was completed just after midnight in the early hours of April 20, the day of the blowout. Later that morning, the drill pipe was pulled out of the well, and the well seal assembly was placed at the wellhead on the bottom of the Gulf floor. At 7:30 a.m., as part of the regular morning operations discussion, it was decided to forego the cement bond log and instead perform positive and negative pressure tests, in accordance with the regulations and the well decision tree.

As part of the tests in lieu of a cement log, the well was pressure tested with mud to 2,700 psi. This test was performed 10-1/2 hours after the cement job had been completed between 10:55 a.m. and noon on April 20, the day of the blowout. The drill pipe was then reinserted into the well to displace mud. The mud logger on the M/V Damon Bankston informed the drill operator that the displaced mud would not be monitored during mud offloading.

The hours and days during which a well is finished and closed are hectic. Many workers are preparing the rig to move on to the next site.

Others are cleaning up their operations just as cementers are doing the final bottom cement, washing up excess materials, and monitoring the job. This wide variety of jobs being performed simultaneously as the well is completed can interfere with the job of the mud logger to watch and measure displaced mud as it flowed into the mud ship. The various activities necessary to clean up after the bottom cementing shared the same ship's hold that was used to contain and measure the mud emanating from the well hole as it was flushed clean in preparation for capping. While the mud logger would normally monitor mud travel as the mud is displaced with seawater, the drill operator instead instructed the mud logger to suspend logging while mud in the drill pipe was displaced with seawater. The drill operator told the mud logger that the he would be notified when the displacement procedure was completed.

Meanwhile, sea water and displacer fluid was pumped into various lines at the wellhead to force out mud. The crew chose to displace mud with seawater because the seawater can produce a cleaner surface for bonding of the cement plug that will offer a redundant seal to close down the well. Phillip Johnson, a professor of petroleum engineering at the University of Alabama, was quoted as saying that such a procedure is not uncommon when normal pressure readings indicate that the well is sealed at the bottom, to ensure a good cement bond for the plug. "But without a good pressure test, it would be reckless to displace," Professor Johnson stated.[91]

BP's choice to flush the mud from the riser with seawater in preparation for a cleaner cement plug ultimately reduced well safeguards. Once the heavy mud is flushed with the lighter seawater, the riser pipe is potentially unbalanced. Only the maintenance of pressure on the seawater can compensate for the displacement of the heavier mud. Of course, if the cementing job and shoe at the bottom of the well were functioning as designed, as verified by a properly interpreted pressure test, such a temporary condition can be tolerated until a cement plug is placed. However, if the well is not properly secured at the bottom, and if the seawater pressure cannot be properly maintained, the level of risk of a blowout rises substantially.

Any losses in a positive pressure test, and the running of a negative pressure test, exacerbate this risk. In addition, throughout these processes, the blowout preventer is held open. This process left high density displacer fluid in the pipe above the blowout preventer that capped the well. Also, seawater pressurized to 1,200 psi was left trapped in a line feeding the wellhead.

The well was then ready for a negative pressure test. In such a test, the well is sealed at the top, and pressure is bled out of the production casing until the pressure is less than the hydrocarbon pressure at the bottom of the well. This negative pressure test is designed to determine if pressurized hydrocarbons can penetrate the cement seals and flapper valves at the bottom of the well.

However, the seal at the top of the well, called an annular preventer, did not seal the well adequately. Hydraulic pressure to the seal was increased until a seal was established. Pressure in the production casing was then reduced. At the reduced pressure, some compression of the fluids in the production casing should have resulted in the movement of 3.5 barrels of fluid out of the production casing. Instead, 15 barrels flowed out. This excess flow should have indicated to the crew that hydrocarbons were flowing past the two barriers at the bottom of the well.

The investigative team observed that this excessive flow may have been missed because the team was observing outflow of fluids from the drilling line rather than from the procedures-specified kill line. Once this discrepancy from BP procedures was detected, the rig crew began to monitor flows from the kill line. Its team opened the kill line, the team estimated that another 3 barrels to 15 barrels of seawater flowed out. The kill line was then closed.

At this point, between 6:00 p.m. and 6:35 p.m., pressure in the drill pipe increased from 50 psi to 1,400 psi. To reestablish the negative pressure test, the kill line was reopened, and .2 barrels, or about 8 gallons, of fluid flowed out. No further flow was observed for the next thirty minutes.

However, the drill team debated why there would be pressure of 1,400 psi in the drill line for a negative pressure test. Witnesses reported that the pressure was due to something the tool pusher had seen on previous wells, which he called a bladder effect. The well site leaders and crew discussed this theory and accepted the explanation for a higher-than-expected pressure at the drill pipe, even though no pressure was encountered at the kill line specified to be used to conduct the negative pressure test. This could have occurred if the kill line had been plugged or a valve in the kill line had been left closed inadvertently.

The investigators nonetheless concluded that the bleeding off of an excessive amount of fluid from the well should have indicated to the well team that hydrocarbon infiltration had occurred. Guidelines for a negative pressure test should have been more closely followed, and the maximum amount of fluids when bleeding off is performed should have been specified. At 7:55 p.m., the negative pressure test was deemed complete. In retrospect, the crew erroneously believed the negative pressure test had been met, even though the well did not have integrity.

4) Penetration of gas was not detected until gas was present in the riser.

The investigators noted the obvious. The primary and overriding role of the drill team is to maintain well control at all times. Even though the team had falsely determined that the well had integrity, the team should have remained vigilant in its effort to detect and address uncontrolled hydrocarbon infiltration that could lead to a well blowout.

As the crew began to prepare to seal the well, it attempted to pressurize the well to return it to an overbalanced condition. The annular preventer seal at the top of the well was reopened, and seawater was pumped into the production casing so that mud in the riser connecting the wellhead to the platform above could be displaced in preparation for capping the well.

However, seawater is lighter than the mud it was displacing. If the well was adequately sealed at the bottom, as the team had incorrectly assumed, the resulting underbalancing of pressures above the well bottom with pressures below would not be problematic. By 8:52 p.m., a sufficient amount of relatively light seawater had displaced enough of the heavier mud to permit hydrocarbon inflow from the failed cement, shoe track, and flapper valves.

The resulting underbalancing went undetected on the platform, as did the outflow of an estimated 39 barrels of fluids by 9:08 p.m. This is known now because the data provided to the computer screens and gauges on the platform was also provided to a second Sperry-Sun computer. This computer readout could be observed anywhere in the world with the authorized Internet connection. However, those on the ship did not notice a sudden drop, and then rise of pipe pressure. Had they noticed, they would have seen an unstable condition arising that would likely telegraph an impending blowout. In the myriad events of the moment, the troubling data went unnoticed. Instead, other operations resumed that would mask this important and overlooked data. Indeed, the crew would miss other spikes in pressure over the next thirty minutes.

Another problem was that the driller did not notify the mud logger on the M/V Damon Bankston that monitoring of fluid outflows should resume. Consequently, the symptoms of mud returns from hydrocarbon infiltration went undetected. The platform crew may also have already been preparing for setting of a cement plug in the casing and fitting a locking collar and may have been distracted.

By 9:08 p.m., the spacer fluid that had originally topped off the riser at the wellhead reached the top of the riser. At that point, the procedure is to shut down the displacement pumping operation, discharge the spacer fluid overboard, and check for oil sheen.

Even if the mud logger was actively monitoring outflow, the flow would stop. Nonetheless, the high pipe pressures would still be observable both at the mud logger's console on the M/V Damon Bankston and the driller's console on the Deepwater Horizon.

As the sheen test was conducted with mud pumps shut down and seawater displacement ceased, pressure in the production casing and the drill pipe continued to rise.

The sheen test was concluded by 9:14 p.m., and the mud pumps began again to recommence seawater displacement of the mud. We now know that an estimated 300 barrels of fluid would have been ejected overboard by an equal amount of hydrocarbon infiltration from below. The pumps were turned off once more at 9:31 p.m. The well continued to unload oil at a rate estimated to be between 60 and 70 barrels per minute. From 9:31 p.m. to 9:34 p.m., the pipe pressure rose another 560 psi, and the drilling team first began to discuss unexpectedly high pressures. By 9:38 p.m., hydrocarbons had entered the risers and were making their way from the wellhead to the platform as the operators remained unaware of impending dangers.

Not until 9:41 p.m., 43 minutes after the first indications of excessive pressures would have been available at the mud logger's and the driller's consoles, did the rig crew first respond to an increasingly dangerous and out-of-control well situation.

5) The drilling team failed to take actions that could regain control of the well.

Regardless of the distractions that might have prevented the drill team from first noting a destabilized well, their role is to maintain well control at all times. By 9:40 p.m., mud was beginning to flow uncontrolled on the deck of the platform. At 9:41 p.m., the crew attempted to shut down the annular preventer valve at the wellhead. It was the same valve that did not effectively shut down on first try earlier, in the negative pressure test procedure.

The crew's emergency procedures had not prepared them for such a high flow, and escalating, runaway event. However, this condition was not unfamiliar to Transocean. One of Transocean's sister rigs had encountered a similar condition just a few months earlier. On December 23, 2009, gas entered a rig in the North Sea that was also performing a seawater flush toward completion. That crew, too, had only one well-bottom safeguard in effect as they performed a seawater flush and negative pressure test. However, when pressures began to rise quickly, indicated hydrocarbon infiltration, the crew was successful in shutting down the well.

In response to this accident, Transocean produced a PowerPoint presentation so that others in the company could learn from their near disaster. It outlined how to deal with an imbalanced well condition, just as would also occur with the Deepwater Horizon crew. This mandatory action, in the form of a document labeled "Lack of well control preparedness during completion stage" was released company-wide on April 14, 2010. The Deepwater Horizon blowout occurred six days later. There is no indication that the Deepwater Horizon crew saw the mandatory action.[92]

6) Improper routing of gas-laden mud created a hazardous condition.

As the crew grappled with a runaway well, hydrocarbons were already in the riser and rapidly travelling toward the platform. Surmising that the mud spewing from the drill tube may be laden with hydrocarbons, the crew diverted the riser flow toward their Mud-Gas Separator (MGS). However, the MGS was not designed to separate the high volume of gases and flow of mud now rising uncontrollably from the runaway well. By 9:47 p.m., the pressure in the drill pipe had rose from 1,200 psi to 5,730 psi in one minute.

With the MGS overwhelmed, it began to vent excess gas flow through a twelve inch gas line. This vent, on the top of the platform, directed what had become a very rapid flow of gas downward onto the platform. Had the crew instead diverted the mudflow overboard, the gas would likely have dissipated safely. The team was concerned, however, that such a diversion of mud overboard would create a reportable violation of EPA regulations. Consequently, the mud, oil, and gas in explosive concentrations were vented in the direction of the heating, ventilation, and air conditioning (HVAC) ducts, and into confined spaces around and underneath the platform. This decision, possibly motivated to avoid a minor EPA violation, instead created the largest offshore blowout-related spill in history.

7) The rig gas and fire system did not prevent gas explosion.

Some parts of a drilling platform are designed to operate without explosion even in the presence of hydrocarbons. These "electrically classified" areas are designed to be spark-free, to prevent hydrocarbon ignition and explosion, and have both alarms and automatic mechanisms to protect the crew in the event of hydrocarbon intrusion.

Areas of the platform that are of lower risk and maintain a greater isolation from areas in which hydrocarbons could infiltrate have lower levels of protection. For instance, the fans in the HVAC system are not designed to automatically turn off in the event of a hydrocarbon alarm. The investigators believed that, in the case of an emergency, some areas are not automatically turned off so they can maintain critical platform functions, such as the dynamic stabilizers that keep the rig positioned, remain powered.

Such a design failure is not problematic so long as hydrocarbon flows do not find their way into areas with spark and ignition. In this instance, hydrocarbons unexpectedly found their way into the engine rooms, causing at least one engine to overspeed. Witnesses noted the sound of a runaway engine. Brightening rig lights seem to confirm that the runaway engine caused a generator to speed out of control. The possibility of an electrical fire or spark under such conditions increased the likelihood of a natural gas explosion.

This explosive condition was in an area of the platform without the detection and explosion prevention mechanisms found on parts of the platform deemed more vulnerable. There were also reports that gas alarms had been routinely turned off by Transocean's Deepwater Horizon crew.[93] At 9:56 p.m., the first explosion rocked the platform.

8) The BOP fail-safe device multiple redundancies all failed to shut down the well.

At this point, seven potential barriers to the Deepwater Horizon explosion had failed due to cementing failure, equipment failure, human error, and a failure of natural gas detection and explosion prevention mechanisms. The final fail-safe device is aptly named a blowout preventer (BOP).

The BOP is a large and heavy apparatus that stands atop the wellhead. Weighing 300 tons, the BOP on this wellhead was designed and built by Cameron International. Called "tweezers" or "pincers" by drilling crews, these BOPs use three hydraulic rams to pinch the large riser pipe shut in the case of a runaway well. The design uses three rams because it is possible that one ram could, coincidentally, try to pinch shut the pipe in the location of one of the occasional joints between two drill pipes. In such a case, there remain one ram to pinch, and another to act as a redundancy.

The crew of the Deepwater Horizon attempted to initiate the ram sequence seven minutes after the explosive mix of natural gas caused the first explosion on the platform. However, the emergency switch did not activate the blind shear ram, perhaps because of damage caused to the electrical system on the platform as a consequence of the explosion.

In the event of such a communications breakdown, the BOP will go into automatic mode. This mode is tripped if there is a communications failure with the platform, but also requires a failure of hydraulic power. There was an indication on the platform of loss of hydraulic pressure. In such conditions, the automatic mode function (AMF) should have been initiated. If so, the ram would pinch down the well when either the blue or the yellow pods, two redundant ram systems that can work independently of the other, would complete the shutdown sequence.

Neither pods managed to shut down the well.

Finally, the BOP has a manual activation valve that can be controlled by a Remotely Operated Vehicle (ROV). Within 33 hours of the explosion, an ROV successfully operated the blind shear valve. The valve failed to activate the blind shear.

After the accident, the controls to the blue and yellow pods were retrieved from the BOP. The battery on the blue pod was found to be discharged. A critical solenoid on the yellow pod was also found to be defective. It was also speculated that debris in the pipe, insufficient or improperly routed hydraulic pressure, or a defective seal, caused the manual shutdown ram to fail.

The BOP was removed from the sealed well on September 2. The Department of Justice took immediate possession of the BOP. There have been charges that the BOP was leaking fluid before the disaster, which might prevent the rams from functioning. Transocean has been accused of failure to maintain the BOP, as it has been cited on at least one previous occasion. Transocean had also admitted that it had departed from the MMS-mandated periodic disassembly and recertification of the BOP. Instead, it had redesigned the BOP in a way that replumbed a critical ram to act as a test ram instead of a ram that could close down the pipe in an emergency. This replumbing of the BOP may have explained why ROVs sent to the ocean bottom shortly after the explosion and spill were unable to initiate the BOP even as it successfully operated the valve that would normally shut down the well. And, BP has been accused of modifying the BOP and of commissioning repairs and maintenance to the BOP in facilities unauthorized by the manufacturer, Cameron International. It is not yet known if any of these accusations, if true, led to the failure of the BOP. The forensic investigation will hopefully identify ways to prevent such a BOP failure in the future.

The BP investigation team notes that each of these eight barriers to catastrophe had to have failed to cause the runaway well to create the conditions that would lead to the largest offshore oil spill in U.S. history. The system was redundant, seven times over, and, yet, the system failed.

At no other time in oil drilling history has such a multiple failure existed. There have been 14,000 deepwater wells drilled without this outcome. A complete investigation into the Deepwater Horizon accident should result in procedures, regulations, and practices that prevent such an event from ever occurring again.

15
The Principal–Agent Problem and Transocean

Corporate responsibility lies in three realms – the economic, the ethical, and the legal. Much of the posturing between half a dozen entities, BP, Transocean, Halliburton, Cameron International, the regulatory agencies, and even first responders, are all based on fears of legal liability. However, there are greater principles invoked by the disaster and the responsibility of various interested parties who could have helped avoid it.

In an era of large corporations and specialized contractors, it is impossible for an entity like BP to control all facets of its organization. Instead, modern organizations, as the "principal" must manage the efforts of contractors, its "agents." This principal–agent problem creates another set of challenges. The principal must work to align its interests, or the interests of its shareholders, with the interests of the agents. There are a variety of solutions to this "principal–agent" problem. However, no solution perfectly aligns the goals of the principal and the agent. Tensions and diverging incentives invariably creep in.

Even the goals of the principal cannot be described simply and definitively. The advantage of limited liability companies, such as BP, Transocean, Halliburton, Cameron International, and others, is that they offer shareholders an opportunity to invest in their enterprise without risking any of the shareholders assets but their initial investment. Should a company become illiquid, meaning the value of its assets are less than the value of its liabilities, the company must reorganize or dissolve. The investment of its shareholders is lost, and its creditors may even receive only partial value once the assets are liquidated or reorganized.

Limited liability

The limited liability company is attractive to investors because of the liability protection it affords. Investors can share in all the gains of a company without taking a personal stake in any losses beyond the shareholder investment. Consequently, a wide diversity of investors will own the shares of a company such as BP.

This diversity of shareholders cannot agree on the same diversity of values that each investor may maintain. Instead, shareholders are united in agreement that the limited liability company should be profitable. While each individual investor might value the environment, their family, political causes, their faith, or anything else of individual importance, they unite on the attainment of profit by the company in which they invest. This myopic pursuit of profits is the motivation of the firm. The firm's board of directors will attempt to balance short-term profits with long-term profits to offer its shareholders some cash, or dividends, each year, and a higher corporate value, or capital gains, for the future.

The simple application of principal–agent theory, with shareholders as principals, and the management and employees of the corporation as agents, does not imply that there could not be corporate values in addition to the values of the principal shareholders. Indeed, when BP changed its name from British Petroleum to BP, shortly after its merger with Amoco as a way to create a stronger presence in the United States, it began to associate its moniker with Beyond Petroleum. Its logo emphasized the color green, to reflect green values, and yellow, associated with energy, like the sun. BP cultivated a reputation as a corporation concerned with our energy future and our environment. BP pursued this corporate philosophy not just because it believed these values were shared by its investors, but also because these values would bode success for the corporation, in its culture, pursuits, strategic plan, and marketing.

One could be cynical about an environmental corporate value from an oil company. Certainly, a major environmental calamity can cast into doubt all such efforts. However, at some level, it does not make sense to challenge too deeply any corporate value. While corporations may try to convince the public, with various degrees of success, that they are dedicated to the environment, safety, the consumer, or the planet, they are ultimately responsible to their shareholders. Any strategies to which they profess must be consistent with shareholder value. And, just as we saw with Exxon in the aftermath of the Exxon Valdez disaster, the public relations branch of a modern corporation may be advocating for one corporate position while the legal team is pursuing another seemingly inconsistent strategy.

This conclusion is not to cast a negative light on corporations. Rather, it is in the nature of a corporation to protect and enhance its profitability, just as parents might be motivated to protect their child. Each does so with dedication and one-mindedness. It is meaningless to try to determine what motivates the corporation, the parent, the evangelical, or the politician. One cannot really know. Instead, each can only be judged by the pattern of their actions and the success of their strategies on behalf of those they represent.

Consequently, we should not be surprised if an oil company attempts to clean up its spills on the one hand, but also tries to reduce its legal liability, on behalf of its shareholders, on the other hand. Likewise, we should take with a grain of salt the efforts of each of the principals involved in the Deepwater Horizon spill to try to shift liability on the other principals.

BP is the agent that acts on behalf of its principals, the shareholders. It is also the principal that acts to try to align the efforts of its agents, contracting corporations such as Transocean and Halliburton, on behalf of BP. In turn, Transocean is acting as an agent for BP, but also an agent for its shareholders. These relationships are necessarily complex, governed not by contracts alone, but also by the myriad decisions each agent makes, and on the efforts of their principals to align agents' interests with the interests of the principal.

The company man

In the parlance of the oil and gas exploration industry, the context within which this case study is confined, the oil company is considered the responsible party. The oil company representative on a drilling platform is known as "the company man." While all understand that the company man is paying the bills, he delegates his responsibility to the platform captain, the drill manager, or the tool pusher responsible for running the drill pipe, at various times.

The company man wants to produce a successful well on budget and on time. As the residual claimant that earns a profit equal to the amount of revenue generated by a successful well, net of costs, the responsible party earns a reward by taking on an acceptable level of financial risk. Each decision the company man makes is benchmarked against such a risk–reward tradeoff. Even the risks any principal must take in its pursuit of reward must be translated into financial terms. The company man is inevitably tied to the financial risk–reward tradeoff, as agent to a corporation that must translate every decision into such a metric.

The corporate contractors respond to a different set of incentives. Their reward is that they are paid by the day. The company man becomes most anxious if a prospect is over budget and past its estimated timeline. The drilling contractor internalizes this same pressure only through the best efforts of the company man to heighten its sense of urgency, and in its desire to maintain a reputation for completing contracts on time and on budget.

On the other hand, if the contractor's reward is primarily in the fixed daily contract price for a platform and its crew, approximately $500,000 per day in the case of the Deepwater Horizon, it maximizes its net reward by reducing its costs and by taking on less risk. In essence, each party to the contract attempts to maximize its net reward while it shifts as much of its risk onto the other party.

In recognition that, in the event of calamity, the deepest pockets will certainly be held accountable first, the oil company realizes that the risk it ultimately takes on, in a legal sense, is large. Regardless of whose decisions might have led to calamity, all will naturally look to the oil company as the responsible party. Given the immense number of decisions that must be made for an enterprise employing almost 200 workers for up to six months, it is impossible for the company man to control all variables. Instead, the art is in delegating authority, and the humanity is in living with the consequences.

While this contradictory system of incentives functions by having the company man constantly emphasize profitability and speed, and the contractors emphasize crew cohesion and a sustainable pace, everybody well understands this dance. Tensions arise, but similar disagreements occur on every contracted platform most every day, to some degree. Each side of the principal–agent relationship plays their part, and they move on, as professionals. At the regular morning meetings, all understand that the company man makes the critical decisions, but these decisions are enlightened by vigorous dialogs and, at times, heated arguments, on an ongoing basis. At other times, professionals make critical decisions consistent with their training and their understanding of the facts as they present themselves.

All the while, the company man will talk about production goals and about the number of days they are behind schedule, and the contractor will emphasize that they are moving at the best pace possible. This creative tension may actually stimulate debate over the balance between reward and risk that biases decisions in the direction of greater safety. Ironically, an oil company that also owns and manages the platform may not benefit from that constant check, or the same ongoing tension.

When a well is successfully completed, all share in that reward. The oil company brings into production a profitable well. The drilling contractor was able to complete a job that is often deeper, more technical, or more challenging than the one before. And, all move on to the next project.

There is one interesting aspect of this dance of divergent interests. All members of the crew are in equal peril if a deep-sea well blows out. Given this shared vulnerability, any crew member has the authority to sound an alarm and halt the operation if she or he believes the platform and crew is in peril. This authority is not taken lightly, but the responsibility to others is stronger still.

A bureaucracy too big to manage?

One of the conclusions drawn early on in the various investigations was whether one of the largest companies in the world was simply too big to oversee its various global operations. Indeed, organizational challenges become difficult as additional layers of management separate operations in a rig on the Gulf of Mexico from discussions in a board room in London.

However, modern organizational theory has learned to adapt to these complications by decentralizing decision-making down to lower levels that are better able to understand the problem, even if they are unable to absorb the ramifications of a catastrophe. Much has been made of whether BP headquarters in London was monitoring the Deepwater Horizon rig on a daily basis. More likely, decisions were made at the level of the exploration division in Houston and on the rig itself. While these decisions proved fateful, this overall organizational structure is reasonably robust. After all, before this accident, the major oil companies had accident, spill, and fatality rates orders of magnitude lower than the industry average, once the size of operations are taken into account.

16
The Management of Risk

The principal–agent tension between the company man and the contractor may actually create a healthy dialog that biases decisions toward safety. However, decisions must inevitably be made that provide some balance between reward and risk.

It would be compelling to argue that an entity should not take on risk, for any potential reward. However, this is untenable. For instance, we know from crash data that one's chances of survival in an automobile collision are higher in an expensive well-engineered car with a 5-star crash rating. Only 5 of 217 cars in model year 2010 in the United States had a 5-star rating in all 6 safety categories. Yet, other riskier vehicles remain popular. Obviously, we all are willing to sacrifice safety, and increase our driving risk, for other rewards, such as cost savings.

We can even ensure that there is never again a major deep-sea well accident. We can ban offshore drilling altogether. However, it is estimated that upwards of 60% of oil production now comes from offshore wells.[94] Conventional onshore oil cannot make up the difference. Nor will consumers tolerate $300 per barrel oil and $10 per gallon gasoline. Until national economies transition to sustainable energy sources, offshore oil risks will have to be managed.

The science of risk

Risk management is an elaboration of a pair of basic economic principles. When one pursues a course of action, the activities with the greatest benefits and lowest costs should be pursued first. And, one should stop pursuing activities when the benefits of the last activity equal its costs.

These premises assume that all costs are included in the analysis, including the cost of risk. To compare benefits to the cost of a risky decision, a decision maker must calculate the impact if a risky event occurs times the probability of the risky event:

Effective cost of risk = Damage caused by risky event x Probability of the risky event

When there are a number n of possible scenarios that can go wrong, each with a different probability P_i, and each with a different level of damages D_i, the effective cost of risk must be summed over all the various contingencies:

$$\text{Effective cost of risk} = \sum_{i=0}^{i=n} P_i D_i$$

For instance, in the case of a catastrophic well blowout, the damages described in this risk analysis include the cost of injuries and lives lost, the property damage and lost revenue from a destroyed platform and drilling infrastructure, the opportunity cost of revenue lost or delayed because the platform and the well are unavailable, the cost of repairing the well damage, the physical and environmental costs of the oil spill, the economic costs of those that suffer displacement because of the accident, and the long-term financial costs to the company and the industry.

While it may seem ethically bereft to make calculations based on the probability and value of a human life, such calculations are inevitable. Just as we deduced that we each calculate the value of the lives of our family members when we choose a vehicle of less than ideal safety, engineers must often build to minimize, but never completely eliminate, risk. While we are all especially conscious of the risk of the loss of human life, the prevention of such a risk would require humans to forego activities we engage in on a daily basis and at an affordable cost.

These costs vary depending on the catastrophe scenarios. We have noted that there are a large number of potential scenarios, depending on which barriers to a blowout are breached.

For instance, in the risk management equation, the assessor must calculate the likelihood of a cement bond failure and the cost of detecting and repairing it. The assessor must compare this to the benefits of the repair, in greater well reliability.

However, the analysis does not stop there. A failed cement job increases the probability of a blowout because a well-functioning shoe track becomes more critical to blowout prevention.

While the probability of cement failure is quite low, it remains the leading cause of well blowouts. In an exploratory well, the well is protected from this cement failure through the shoe track and subsequent barriers, plugs, and seals higher in the well. A blowout in an exploratory well that has been capped awaiting later production is a very low probability event because it requires all these barriers to fail simultaneously.

For illustrative purposes only, let us assume the following probabilities:

Probability P_c of a catastrophic well bottom cement bond failure $=.1$
Probability P_s of a shoe track failure $=.05$
Probability P_b of a blowout preventer failure $=.1$
Probability P_a of an annular seal failure $=.1$
Probability P_p of a cement plug failure $=.05$

The probability of all these events occurring would be the product of each individual probability, assuming that these probabilities are independent:

Probability of a catastrophic well blowout
$$= P_c * P_s * P_b * P_a * P_p$$
$$=.1*.05*.1*.1*.05$$
$$=.000025$$

In other words, a blowout that must breach all five of these fail-safe devices will occur once for every 40,000 wells drilled.

Actually, real world blowouts are less frequent than one in 40,000 wells drilled. This tells us that these individual systems have a lower-than-assumed probability of being breached. Our example demonstrates the dramatic reductions in risk when multiple redundancies are employed. For the Deepwater Horizon blowout to have occurred, at least eight barriers were breached. There remains discussion about whether there should have been an even greater number of barriers. For instance, a greater number of annular seals in the gap between the well lining and production casing would add additional barriers to possible pathways for flow of hydrocarbons from the bottom to the top of the well.

However, some of these redundancies may have reduced the probability of a well blowout under other scenarios, but would not have affected the likely pathway that this blowout followed. Evidence suggests that hydrocarbons migrated up the interior of the production casing, not the annulus between the casing and the well lining. Consequently, an emphasis on the creation of risk through practices not related to a particular catastrophe is irrelevant.

In other words, the ex-ante management of risk, meaning the economic decisions that are made in the design and the implementation of the well design, must be distinguished from the ex-poste analysis of risk.

For instance, consider two questions. If a blowout preventer is the only barrier that can protect a well on the verge of blowout, what investment in blowout preventer technologies is warranted? This is a very different question than the determination of an optimal blowout preventer that would only come into service if all other barriers fail. The first scenario loads all the potential costs of a catastrophe on one barrier, while the second scenario has the barrier as one of a system of redundant barriers.

Let us use the probabilities from our previous example, and assume that the cost of a catastrophic oil spill of the Deepwater Horizon variety was at the upper end of estimates, including maximum fines and loss of corporate good will, of $40 billion. Let us also assume that the cost of making a blowout preventer that is twice as reliable is one million dollars.

In the ex-ante risk management analysis, the expected cost of a catastrophic oil spill is given by the probability of such a spill times the damage that will occur:

$$P*D = .000025 \times 40,000,000,000 = \$1,000,000$$

Using these illustrative numbers, halving the probability of a catastrophic failure, ex-ante, would then save $500,000. Ex-ante, it would not be economical to invest more than $500,000 in greater blowout preventer reliability, even for the largest single environmental catastrophe the industry has ever seen.

However, if the question is asked in a different way, the conclusion is very different. We can calculate the ex-poste optimal investment in a doubling of blowout preventer reliability, given that all previous barriers have failed. Such a doubling of reliability would reduce the expected damages from $4 billion to $2 billion. In other words, under the ex-poste analysis, the blowout preventer improvement should have been worth $2 billion. Most would agree that to invest such an amount in blowout preventers within a system of redundancies would be uneconomic and would do little to mitigate overall risk.

Risk management must be measured against the probability of all possible scenarios. Unfortunately, all too often, the failure will be judged from an ex-poste perspective.

For instance, the investigation in the Space Shuttle Challenger in-flight explosion concluded that a relatively inexpensive O-ring failure caused an accident that set the space program back billions of dollars and many years. This ex-poste observation does not suggest a billion dollar O-ring. In the final analysis, it may be the case that the O-ring was adequately, if not perfectly, engineered, but that there should be changes made in other designs further up in the redundancy chain.

Such analyses are very complicated because they require a great deal of data on the costs of various safety technologies, estimates of many damage scenarios, and the attribution of risk to each factor that must be managed. Computers and software are employed to manage these risks in modern enterprises. For instance, "bow-tie" software can illustrate the potential risks that can occur as redundant barriers are breached and the overall risk is escalated.[95]

However, when circumstances necessitate a spot decision, engineers must tinker with the design based on their instinct or calculations of the risk–reward tradeoff.

Another example of this confusion between ex-ante and ex-poste risk often occurs in the media following a major catastrophe. After a catastrophe, there is inevitably some whistle-blower that had called for changes at the unfortunate facility. The concerns of this whistle-blower are instantly given greater credibility, perhaps warranted, but perhaps unwarranted.

We also see this phenomenon in the broader economy. At any given time, there is surely an economic prophet who claims that global financial market will experience a major crash. There are many such prophets, each stating a different period for the crash. When the crash does occur, one of these prophets, who perhaps happened to be in the correct place at the correct time, becomes an instant genius. However, it is unlikely that the prophet will be successful again, despite his newfound celebrity status.

This differentiation between ex-ante and ex-poste prognostication should not be entirely dismissed, however. Often, whistle-blowers are attempting to point out circumstances that defy easy quantification. The software programs that manage risk, and the equations that balance risk, require the user to be able to measure the probabilities, costs, and benefits of various scenarios, and estimate how various factors influence every other factor.

Such attributions are likely accurate for events that occur regularly and generate sufficient quantitative data for subsequent analyses. However, risk management is very difficult when it tries to model events that have never before occurred or which occur infrequently and randomly. It turns out that the attribution of risk in such events is very unreliable.

To see this, let us describe some artifacts of the Poisson process, a statistical model that is used to estimate the time of arrival of an event that occurs at a regular rate, even if infrequently. An example of this might be the modeling of a hundred-year flood.

Let us assume there is no historical flood data available and one observes a flood in the first year. One might conclude that the flood occurs once every year. In fact, a flood event in that first year is a one-in-a-hundred year event, or .01 probability, rather than the 100% annual probability one might surmise, given an absence of historical data.

I am reminded of a scene in the John Irving novel and movie *The World According to Garp*. In the scene, Garp is contemplating buying a farm. While inspecting the farm, he observes an airplane fly into the barn. While others might incorrectly consider the property disaster-prone, based on the ex-poste assessment of risk, Garp declared that the farmhouse to be the perfect property to buy because it was then "disaster-proofed." After all, what is the probability (ex-ante) of such a rare event ever occurring again?

Obviously, this distortion in data perceptions can occur in either direction. An unlucky spat of failures that can occur both randomly and systemically, can be deemed to occur with greater regularity than justified. On the other hand, an event that occurs infrequently can cause risk managers to underestimate the true risk. In other words, success can contribute to safety complacency.

Recognizing that nonexistent or anecdotal data can distort the assessment of risk, managers can use a technique called Enterprise Risk Management (ERM). This technique uses the same sort of sequential iterations that quantitative risk management software employs. However, it also introduces a greater reliance on risk self-assessment across the enterprise, even if such self-assessments are difficult to quantify. The technique fully employs the entire organization in the assessment of risk, in total, not isolation. It is able to model the true organizational cost of risk, not merely the cost to one entity within the organization. It also models how risks compound over time, and models the highest long-term risks as the first priority to remediate.

Risk management in offshore drilling is guided by the principle of safety case, in which potential risks are outlined, and measures and responsible parties are identified to mitigate the risk. For instance, a heavily redacted document investigated by the Government of Australia following a well blowout at the Montara Wellhead Platform on August 21, 2009 in the Timor Sea described the various processes and the overlap of oversight responsibility for the cementing of the well bottom. The safety case had to be approved by the regulatory agency, and the actions had to be performed, monitored, and signed off by the responsible parties and overseers.[96] The safety case acts as a complete risk assessment map for the drill plan, with fully defined responsibilities for each party. In the Montara safety case, the full unredacted results of the investigation have been embargoed, under speculation that there may be prosecution of the cement contractor, Halliburton International, or other contractors.[97]

Drilling companies follow international safety and risk standards as formulated by the International Standards Organization (ISO) in ISO 17776:2000.[98] These standards are augmented by risk management assessment guidelines created by the International Association of Drilling Contractors and by government regulators.

However, these protocols also assume that there is an integration of engineering and management. Obviously, in the Deepwater Horizon operation, interdepartmental and inter-corporate communications were poor, perhaps because of a long history of past exploration success that bred complacency. The widespread use of contracting and the process of decentralization within organizations may lead to greater cost efficiencies which translate into greater corporate reward. They also create greater potential for management risk that can exacerbate the more familiar and much better understood engineering risks.

Best communications practices to reduce risk

Effective risk management also requires successful communication of risk within the organization and to the public. Based on work by Covello and Allen,[99] the U.S. Environmental Protection Agency has published "Seven Cardinal Rules of Risk Communication."[100] These rules specify that an entity should:

1. Accept and involve the public as a legitimate partner.
2. Plan carefully and evaluate your efforts.
3. Listen to the public's specific concerns.
4. Be honest, frank, and open.
5. Coordinate and collaborate with other credible sources.
6. Meet the needs of the media.
7. Speak clearly and with compassion.

Such industry and professional standards are helpful because they allow organizations within an industry to take advantage of pooled data and best practices. Legally, they also provide a certain liability shield by creating a yardstick by which decisions should be judged and bench-marked based on a *reasonable person standard*.

Perhaps an unintended, but nonetheless beneficial, outcome of the Deepwater Horizon blowout and spill is that it has provided risk data for low probability events. Consequently, we can be confident that every major hydrocarbon explorer and contractor is reevaluating its risk management models. For instance, very soon after the Deepwater Horizon spill, ExxonMobil, ConocoPhillips, Chevron, and Shell announced that they had instigated a quick response spill protocol that will mimic what BP had to put together immediately after the spill.[101] But while BP mounted what may well be the world's most comprehensive environmental remediation effort more quickly than any company before it, no doubt driven, at least partly, by necessity and by external pressure, the consortium of other oil companies that drill in the Gulf of Mexico have stated they shall take one year to assemble a similar effort.

Fortunately, BP recently offered to share all of its resources, investment in spill remediation and blowout repair, and expertise, with other oil companies operating in the Gulf of Mexico. Should the other oil companies accept the offer from the first company to experience, repair, and remediate a spill on a scale previously unknown, it is hoped that the Deepwater Horizon tragedy will be the last such spill.

Part IV
The Spectacle of the Spill

Humans have always maintained an uneasy balance between our need for this precious resource and our stewardship of the environment that has for so long contained it. This balance has recently been placed in a different perspective. We now must face the glaring risks we take. Oil has moved into the realm of the uncomfortable. Just as most people would prefer to avoid confronting how animals are processed to create our food, we find disturbing the realization that oil drilling will, inevitably, create spills on occasion that will drift upon a populated region somewhere in the world.

17
For All the World to See

Obviously, no oil company can afford the stigma and liability of a major oil spill or explosion. The industry would prefer to pump over 20 billion barrels of crude annually without spilling a drop of oil. Even though four million barrels of oil spilled into the Gulf of Mexico is but 1/4000th of one year's production, or less than the share of a single drop in a one gallon bucket, such a spill is by no means the proverbial drop in a bucket.[102]

The attention BP received, as the operator of the Deepwater Horizon platform and crew under contract from Transocean, reflected the American public's interest in the Gulf of Mexico. The public well remembered the bungled rescue from Hurricane Katrina almost five years earlier. However, adding to that perfect storm was a public that recalled the Exxon Valdez spill two decades earlier. Just as one major spill may not a pattern make, the second major spill indicates to the public an emerging pattern of corporate irresponsibility. Finally, compounding this conclusion was a public disgusted by a decade of corporate irresponsibilities, beginning with Enron, and ended with 20 months of economic misery precipitated by a financial crisis induced by companies too big to fail, followed by the public bailout of these same companies.

The public cynicism created by the irresponsible acts of the federal government in New Orleans, memories of oil-soiled water birds in Alaska, and the bailout of corporate stalwarts such as American International Group (AIG) and Bank of America (BofA) only fed into the disgust when BP accepted responsibility for the largest oil spill the Gulf of Mexico has ever seen.

Historically, the media has served a sometimes noble and, at other times, notorious role for the American public. The United States was the first nation to enshrine freedom of the press as a constitutional right. The media allows the public to make some sense of a baffling and increasingly complex modern life. In doing so, the media is, on occasion, the most trusted, and at other times the most reviled institution to which the public regularly subscribes.

The media has evolved with technology. From the world's first pamphlets soon after the invention of the printing press, to the first radio news broadcast in 1920, upon the election of Warren Harding, and the first television broadcast 15 years later as part of Hitler's propaganda machine, the media has evolved with every twist in technology.

In the American model, especially, the media has been driven by its ability to rivet readers, listeners, and viewers so that it can guarantee an engaged public to an interested set of advertisers. There is nothing as riveting as strong, even disturbing, images, which has even given rise to the television news production standard "if it bleeds, it leads."

Television news had cornered the market on visuals. Its ability to pan along an oiled beach, fly over a slick, or zoom in on a struggling bird, added movement and life to images. And, with the creation of cable news with the Cable News Network (CNN) in 1980, these images could be put on a 24-hour loop, and offer any viewer those images that move, at any time the viewer finds convenient, or over and over again for the viewer that seeks constant reinforcement.

The issue is no longer who can produce the better, more thoughtful, and more balanced portrayal of those events that affect our lives. Rather, the industry rewards those that can create and reinforce the most memorable images. News agencies know that humans crave order. Those news providers that can reinforce the public's suspicions and fears can succeed in its goal of securing viewers, and galvanizing advertiser support. And, when the public is unhappy and cynical, the media will follow.

The Pew Research Center for the People and the Press documents with its News Interest Index how the media responds to the interests of the public. In their July 14, 2010 article entitled "Modest Decline in Oil Leak Interest, Sharp Decline in Coverage,"[103] they noted that the American public was very interested in the Deepwater Horizon fire and spill almost immediately after the explosion on April 20, 2010. Within a couple of weeks of the fire and spill, 58% of the surveyed American public reported they followed the events very closely. The next most followed story in early July, as interest and coverage peaked, was followed by only 13% of the population.

With that steady increase in interest, the media began to follow and fuel the passions, and devoted 38% of news coverage to the spill a month after the explosion and fire. Constantly bombarded by BP's own strategy of transparency, influenced perhaps because of the EPA's risk management principles entitled "Seven Cardinal Rules of Risk Communication," the public could view, in real time, oil spewing out of a 16 inch pipe and into the Gulf. Never before had we been able to view a spill, even at its source, in real time, in a populated area, and with 24 hours coverage. This was the first spill of the transparent era, or at least the first spill that was in a location for which the public showed great regard.

As the oil incessantly spewed out of the picture in the corner of the newscast television screen, the camera would scan over slicks, tarred beaches, and blackened turtles. The public was riveted, with a majority of surveyed viewers indicating both that they were following the story more closely than any other, and that the media was offering the right amount of coverage.

The public would remain captivated, with the majority of Americans claiming they followed the story very closely. Then, on July 12, BP successfully capped the well and the images stopped. The media could not continue to loop a spewing pipe when the public knew the oil had stopped. BP had succeeded in cleaning the beaches, and the slicks on the surface of the ocean rapidly dwindled in size and soon disappeared. There were no more visuals, so the media very quickly lost interest. By mid-July, the coverage that dominated more than 40% of news time had fallen into the single digits.[104]

The constant media coverage may have the effect of distorting our collective sense of reality and the decisions that flow from it. For instance, author Barry Glassner, in his book *The Culture of Fear: Why Americans Are Afraid of the Wrong Things*, observed that the amplification of coverage exaggerates our own sense of probability.[105] He used the example that people believe airplane travel is dangerous, even though we are far more likely to be in a fatal accident on the way to an airport than on a commercial flight. We also believe that child molesters are rampant, which cause us to drive our children to school, even though the incidence of kidnapping and child molestation is probably lower than when we walked to school as children.

While anecdotal and irregular data distorts the risk management models toward greater complacency, constant news coverage may actually amplify our perception of risk. We shall see later that this amplification can even be measured in our financial markets. In the United States, though, in the spring of 2010, the media bombardment exaggerated our perception of damage in the Gulf of Mexico, induced us to cancel our tourist reservations to Gulf communities, and likely significantly enhanced the economic costs BP pledged to bear.

Get the message out

A consequence of the Exxon Valdez spill in 1989 was a statutory limit of $75 million on the economic damages associated with the spill. International maritime legal principles have always required the responsible party to clean up and remediate their spills. And, also following the Exxon Valdez spill, a civil fine of $1,000 per barrel spilled, and adjusted periodically for inflation, could be levied for the negligent spilling of oil. In circumstances in which the acts of the responsible party rises to gross negligence, this fine could be elevated to $4,000 per barrel, as a criminal sanction that could not be deducted as an expense from corporate income taxes. In comparison, the statutory limit for economic damages is dwarfed for any spill in excess of 75,000 barrels.

In a most unusual gesture that defies American corporate precedent, BP almost immediately accepted financial responsibility for all legitimate claims arising from the spill. We will delve into the legal implications of this gesture later on. For now though, let us test BP's response against EPA's "Seven Cardinal Rules of Risk Communication."[106]

As you recall, these rules specify that an entity should:

1. Accept and involve the public as a legitimate partner.
2. Plan carefully and evaluate your efforts.
3. Listen to the public's specific concerns.
4. Be honest, frank, and open.
5. Coordinate and collaborate with other credible sources.
6. Meet the needs of the media.
7. Speak clearly and with compassion.

Actually, BP satisfied the EPA risk communications criteria quite well, and perhaps better than any other modern corporate transgressor. It accepted the involvement of the Federal government, and coordinated with Admiral Thad Allen, the retired Coast Guard admiral appointed by the Obama administration to act as the public face of the federal effort. In the first days of the spill and response, there was inadequate planning and follow-through. However, the effort became much more coordinated as the multipronged and multiagency response was developed. BP tried to listen, but was stymied by politics and by a less than deft ear for American sensibilities.

While BP was accused of all sorts of mischief, they insisted their process was frank and open. Surely, the BP corporation lawyers were suggesting strategies more in line with the American corporate culture of deny and delay, but BP did prove to be an effective partner with Retired

Admiral Thad Allen and with Secretary of Energy Steven Chu's science team. There are few better examples of such corporate/public coordination that come to mind.

And, BP spoke carefully, given the legal prudence that must govern its responsibility to its shareholders. BP could not simply tell every claimant what it wanted to hear, especially since many claimants could not produce receipts that would substantiate their losses, in shrimp catches, motel room sales, and similar economic damages. BP was put in the unenviable position of denying claims to such individuals, many of whom were angry, naturally, whether or not they actually incurred the damages they claim. After all, BP was soliciting an impassioned and angry response from a group that would likely be denied any relief at all, had BP not agreed to forego the cap to its legal responsibility.

Instead, BP agreed to use an impartial, but respected, external adjudicator, Special Master Kenneth Feinberg, to administer claims for economic damages.

However, BP was less than deft in managing the needs of the media. Its failure was not an issue of denied media access. Indeed, early on, BP had made available real-time videos of the actions of its Remotely Operated Vehicles as it tried to disentangle, and then repair, the nest of pipes on the ocean floor. And, it appointed Kent Wells, one of BP's senior vice presidents, to brief the media almost daily on subsea repair operations. Meanwhile, Retired Admiral Thad Allen also provided daily briefings and question and answer periods on all issues related to the coordinated response.

BP's public relations gaffes instead portend to cultural differences. Even though BP's absorption of Amoco suddenly made it one of America's largest oil companies, most of the corporate management was British. The difference in sensibilities, or even in accents, explains some early missteps by BP. These missteps ultimately cost BP's Chief Executive Officer, Tony Hayward, his job. It also threatened BP's chairman of the board, Carl-Henric Svanberg, for his unfortunate comment about the "small people" affected by the spill. But, while the public partially accepted his mistaken reference to small businesses from a man for whom English is not his native language, the same public was ruthless with Tony Hayward.

Of course, the mere repetition of the name British Petroleum had negative connotations on two fronts. One was of a foreign company at a time when economic mercantilism was rampant as Americans felt financially threatened, and the British to boot, which invoked memories of economic imperialism from centuries earlier. And the other connotation was of Big Oil, not a term in high regard given high oil and gasoline prices. This term British Petroleum was a double whammy that was used with some frequency, and perhaps not without the intended effect at times, even though the company had changed its official name to BP a decade earlier.

It did not help that the original estimate of flow out of the gushing pipe was tagged by the federal government at an inconceivably low 1,000 barrels, and then, a little later, 5,000 barrels per day. The media erroneously and constantly referred to these estimates as BP's numbers once higher estimates were subsequently refined.

And the confident attitude from Tony Hayward, in the midst of the worst of the spill, did not come across as British reserve combined with American bravado. Rather, it came across as insensitive, especially to the Louisiana ear, more associated with French, rather than English passions.

In contrast, local parish politicians in Louisiana could barely contain their sorrow, and exuded a passion and caring that the region had learned to expect from its politicians, ever since Louisiana's immortal governor Huey Long. Even Barack Obama, the president with the stiffest upper lip since Richard Nixon, was counseled to appear angry and appalled.

In such an environment, Tony Hayward's traditionally understated sensibilities were certainly viewed as insensitive. When he said "I'd like to have my life back," it came across as insincere, rather than a bit of an uncomfortable response, and a poorly timed gesture at understatement, rather than the compassionate "I feel your pain" that President Bill Clinton might once have confided. Most people from Britain, or even this writer, a naturalized American who grew up in Canada and lived for a year in London, recognized Hayward's attempt at understatement as a normal, if not uncomfortable, response by Brits when faced with an uncomfortable reality.

As an example of this characteristically and, at times, unfortunate British understated sensibility, on June 24, 1982, Captain Eric Moody was piloting British Airways Flight 9, a Boeing 747 with 263 passengers and crew on board, over Indonesia when it flew into an undetected volcanic plume. The plume of ash scored its windshield to opacity and glassified all four engines to the point that they all failed. Obviously, the passengers detected that the engines were no longer working. They began to pen their last words to loved ones, when British Captain Moody came on the intercom with perhaps the most understated message ever uttered:

> Ladies and gentlemen, this is your captain speaking. We have a small problem. All four engines have stopped. We are doing our damnedest to get them under control. I trust you are not in too much distress.[107]

The British take an almost nationalistic pride in understatement that may be misconstrued by others. To overcome these misunderstandings, BP committed to an advertising campaign that will likely tally to a hundred million dollars or more. As we shall see later, this campaign likely will not be able to remedy goodwill losses many orders of magnitude higher. Over time, BP will likely be able to remedy its lost goodwill within the bulk of the United States, so long as there are no further incidents. However, 21 years after the Exxon Valdez spill, Exxon's behavior in the aftermath has yet to be forgotten or forgiven. BP can learn from the Exxon experience, and has already demonstrated it is not repeating some of Exxon's mistakes. Repair to BP's reputation in the Gulf will be slow, even after the leak was stopped, the slicks disappear, and the claimants are paid. This remediation of reputation will be slower still if politicians find it opportunistic to tap into the anger for political purposes.

For instance, on September 19th, the day that Retired Admiral Thad Allen declared the well sealed forever, I received a donation solicitation from Truthout, an organization that offers an unconventional slant to media messages. The solicitation was designed to enflame passion from supporters so they would donate in disgust of what the world is coming to. Their solicitation, "BP's Propaganda vs. Unembedded Reality," claimed we are just beginning a decades-long era of ramifications in the Gulf. We shall see. If they prove to be correct, we will remember them as prophets. If they prove to be wrong, their comments will be forgotten. In the meantime, fundraisers obviously hope that we will donate so they can perpetuate a story from which they can profit.

Coast Guard Rear Admiral Mary Landry, commander of the 8th District that includes New Orleans, was frustrated by what she perceived as a media motivated more by headlines than by nuance. In a speech at the Clean Gulf Conference in San Antonio, Texas on October 19, 2010, Landry noted that as media compete for readers, they sometimes raise the level of alarm unnecessarily. She went on to add that any confusion in the media about who was in charge of the coordinated efforts to cap the well and clean the gulf was unfounded. She stated in definitive terms that, contrary to speculation and media reports, the government was in charge.[108]

18
Partners in the Problem

BP was hardly alone in the Deepwater Horizon spill. Anadarko Petroleum and Mitsui Oil Exploration Company are 25% and 10% minority owners of the well, respectively. Transocean, the world's largest offshore drilling contractor, even before it swallowed up the next largest competitor a few years earlier, owned and operated the platform that exploded, and most of the employees on the rig on April 20 worked for Transocean. Halliburton, the world's second largest oil services firm, performed the cement job, and Cameron International had built a blowout preventer that had been since maintained and overhauled by yet other contractors.

All these partners retreated following the explosion and spill, and the Federal government, perhaps almost as resented in the Gulf of Mexico as BP has become, stepped in.

Indeed, the degree to which various parties began tripping over themselves to shift blame and liability became almost comical. Even the other major oil companies operating in the Gulf of Mexico volunteered that what happened to BP could not happen to them – until it was pointed at that the emergency response plan proffered by BP was almost identical to their plans, including references to seal rescue, despite the fact that seals do not live in the Gulf habitat.

Obviously, the other companies were trying to distance themselves from liability that they would prefer BP alone bear. The spectacle induced President Obama to complain:

> I did not appreciate what I considered to be a ridiculous spectacle...Executives of BP and Transocean and Halliburton falling over each other to point the finger of blame at somebody else...The American people cannot have been impressed with that display and I certainly wasn't.[109]

160

Regardless of the corporate posturing, BP accepted its legal status as the "responsible party" almost immediately, even if it could not have yet understood the extent of its potential liability. BP had also been invoicing its partner for cleanup expenses, as provided for in the contract with Anadarko and Mitsui. Anadarko responded, quite predictably, that it would not share in the cleanup of a spill caused by BP's reckless decisions.[110] Regardless, BP CEO Tony Hayward stated that Anadarko's recriminations would not distract BP's focus on stopping the leak and restoring the Gulf.

In this backdrop of lawyerly strategizing, BP appeared in the media to be ill-prepared at first. Very early on, BP had determined that ROVs could not shut off the well by manually actuating the blowout preventer. BP pinned their early hopes on an aggressive strategy to drill a relief well to stop what it perceived at first to be a 1,000 barrel a day spill.

On April 27, one week after the blowout, and five days after the Deepwater Horizon sunk and severed the riser pipe, Tony Hayward believed the most expensive aspect of their spill strategy, the drilling of a relief well, would cost upward of $100 million. Hayward promised journalists that BP would mount the biggest spill response in the history of the industry and the company is able to do so "because we planned for it... We will be judged by our response." Later, in the teleconference interview with journalists, Hayward stated "We are determined to ensure that it does not become a major environmental incident."[111]

A week later, BP had opened up cleaning efforts in Louisiana, Mississippi, Alabama, and Florida, and had employed 2,500 people. Hayward continued to reiterate: "We will be judged by the success we have in dealing with this incident and we are determined to succeed."[112] It had deployed, staged, or procured almost a hundred miles of boom used to collect oil and protect the shoreline, but which was useful only when seas were calm.

BP had even contracted with famed actor Kevin Costner's company which was peddling oil/water separators, and was cooperating with a major global shipper that had modified a former tanker, renamed "A Whale," to intake oil through slits cut at the waterline. These unconventional attempts, while necessary to try, were not instrumental in keeping up with oil that was spewing much more rapidly than BP anticipated.

Certainly, BP and the federal unified response team initially underestimated the scale of the spill. The wellhead was losing oil to the Gulf at a rate much higher than it, or the government scientists, had estimated. And, it was confident that, by August, the relief well that was the standard response to such an event, would be complete. At 1,000 barrels per day, the well would have spilled around 100,000 barrels, or less than half the spill of the Exxon Valdez. At the upward revised estimate of 5,000 barrels per day, the spill would near half a million barrels, or twice as large as the Exxon Valdez, but much smaller than the estimated 3.45 million barrels that spewed from the other major Gulf of Mexico blowout, the Ixtoc I. Clearly, BP believed the spill was of a size of 10% or less than what it would later discover.

These underestimates of potential damage, even if BP acknowledged it was a serious spill, were indicative of a firm unprepared for a spill of a scale never before remediated. It would not be the largest spill in history – almost a century earlier, the Lakeview onshore blowout released more than twice the amount of oil discharged. Nor was it the largest offshore oil spill as was the Gulf Oil spill, released as a consequence of Iraqi hostilities. However, BP was ill-prepared to deal with the largest offshore oil spill for which any corporation had accepted responsibility to remediate.

Even if it were able to cope with the physical aftermath of the spill, a somewhat clumsy and plodding bureaucratic approach did not give the appearance of swiftness. At the same time, President Obama was growing unpopular, accused of being too cerebral when the country was in economic distress and faced recriminations over a bungled federal response to Hurricane Katrina almost five years earlier. Obama commended a more firm and federally guided response. BP would have a partner, whether or not it wanted one. Like the memorable scene from *The Godfather*, the federal government was making BP an offer it could not refuse.

BP and the federal government were strange partners. *The Washington Post* published a poll on June 7 that showed neither party was popular with the American public. Most people thought that the oil spill was a major disaster, and that the federal government should pursue criminal charges against BP. At the same time, 69% of Americans thought the federal government had mishandled the aftermath of the BP oil spill. This lack of confidence, compounded by the memory of managed rescue and repairs after Hurricane Katrina, was not as bad as the 81% of the surveyed population that disapproved of BP's management post spill. And, almost six in ten interviewed thought that both BP and the Federal Government were mishandling the spill six weeks into it.[113]

Meanwhile, Obama had insisted that BP had attempted to "obfuscate the amount of damage that's been done by the company..." even though a federal government agency was coordinating and releasing the spill estimates that later proved too low. The administration's anger resulted in an imposition of a six-month moratorium on new well drilling in the Gulf, the cancellation of a lease sell off the Atlantic coast, and a delay in offshore oil exploration in Alaska.[114]

Obama was also angry with the conduct of the Minerals Management Service (MMS), the regulatory body organized to permit and monitor the oil industry, among other sectors that pay royalties the Federal Government in return for mineral rights. We will treat later the administration's response to MMS' history of regulatory dysfunction and the ensuing reform. In May of 2010, though, the administration recognized that it would appear weak and passive if it did not exert greater control over the coordination of the spill cleanup.

On the other hand, the Oil Pollution Control Act of 1990, passed in the wake of the Exxon Valdez spill, mandated that the responsible party must cover the expenses of the cleanup. If the Federal Government was to commission displaced Louisiana ships to assist in the cleanup, it must have BP standing by its side to contract with the captains and write the checks. And, no entity worldwide had the equipment, or the expertise with the Macondo Prospect well, to expeditiously shut it down. Finally, while the Federal Government would want to be the voice with the public and create the sense it is in control, it did not want to assume the ultimate responsibility for the cleanup, especially if it might fail.

This marriage of convenience was challenging. The Federal Government had to give every appearance of being tough with BP, but at the same time foster an effective partnership with them to accelerate the successful shutting-in of the well. The newly elected British Prime Minister David Cameron even had to squander the little political capital he had built up in the short time he had been in office by negotiating directly with Obama on behalf of BP.

BP was able to create an understanding with Obama, under the promise to cover all economic damages, appoint an external special master to facilitate claims, and conduct asset sales so a down payment to a $20 billion claims, fines, and remediation fund could be set aside. In turn, Obama agreed that BP could continue to generate revenue from the sale of Gulf oil, ostensibly to permit BP the resource flow to fund remediation.

This reversal of philosophy, from tough talk to strategic partner, came from the realization that the instinct to force BP into bankruptcy would only hinder the spill response and remediation. The argument that a strong BP would best support an effective cleanup and the ability to pay any fines levied to the U.S. Treasury carried the day.

The broker of the partnership between BP and the Obama administration was Thad Allen, a retired admiral of the Coast Guard. His honest, direct, well-informed, and homespun style was better received than both the understated delivery of BP company men and the spin-laden messages from the administration. Retired Admiral Allen proved to be an effective communicator through briefings and question and answer sessions with the media each day. He became the official voice of the spill, the cleanup, and the subsea efforts to shut down the well.

Allen was frequently asked to speculate about the politics of the spill and its future repercussions. On September 20, a day after Allen declared the well dead, he participated in an interview with Jim Lehrer on the Public Broadcasting Service (PBS) NewsHour. Allen commented on BP's advance preparation for a spill of this magnitude. He observed that it was clear that the response plan employed by all Big Oil firms never anticipated a spill of this type and magnitude.

However, BP's response to this deep-sea spill has created a body of experience and equipment that will allow the world to stop deep-sea blowouts much more efficiently in the future. BP has offered to share this knowledge and equipment with the four other major oil companies who formed the "Marine Well Containment Company." BP's announcement stated:

Release Date: 20 September 2010

Houston – BP announced today its intent to join the proposed Marine Well Containment Company (MWCC) and to make its underwater well containment equipment available to all oil and gas companies operating in the Gulf of Mexico.

This and other equipment will preserve existing capability for use by the oil and gas industry in the U.S. Gulf of Mexico while Chevron, ConocoPhillips, ExxonMobil and Shell build a system that exceeds current response capabilities. Under the terms of an agreement with the Marine Well Containment System operator ExxonMobil, the sponsor companies' project team will utilize full time BP technical personnel with experience from the Deepwater Horizon response.

"We are pleased to announce our plans to join the Marine Well Containment Company and provide the experience and specialized equipment needed to respond to a deepwater well control incident," said Richard Morrison, BP vice president for Gulf of Mexico operations. "We believe the addition of our recently gained deepwater intervention experience and specialized equipment will be important to the marine well containment system."

We next survey the trials and lessons learned as BP responded to the world's most vexing oil spill with the world's most advanced technologies.

19
Engineering a Solution

The BP spill was not the first deepwater blowout in history. However, it produced the largest daily flow, had the greatest sense of immediacy, commanded the most resources in its solution, and was, by far, the most visible of any similar spill. As a consequence, it demanded the most of the engineers who would be responsible for stopping the flow.

The unique engineering solutions to this blowout served as a laboratory for the development of future failures. In turn, the positive legacy of this spill is in what the industry has learned, in spill response, in blowout avoidance, and in the highly technical aspects of the shutting down of risers spewing oil at depths far below the direct intervention of humans.

Well design

To best understand the solutions, we must summarize the problem.

We have documented the factors that make deepwater exploration unique. The industry considers deepwater as a wellhead depth more than 3,000 feet and ultra deepwater as a depth greater than 7,000 feet. Today, there are examples of exploration at depths greater than 10,000 feet, and production wells over 8,000 feet deep. Oil companies now successfully drill for oil that is over 30,000 feet deep.

One defining characteristic of deepwater exploration is that they are beyond the depth at which divers can function. The Self Contained Underwater Breathing Apparatus (SCUBA) diving gear typically consists of a soft-shell wet or dry suit, mask, tanks filled with a compressed breathable mixture of nitrogen, oxygen, and, at greater depth, helium, and a pressure regulator. In essence, as a diver goes to greater depth and must counteract the increasing pressure of a higher column of water from above, the pressure of the breathing gas must rise proportionately.

If this gas were not at a similar pressure as the pressure of water on the outside of the diver's lungs, the diver would be unable to inhale. The pressure regulator is designed to increase the pressure of breathing gases to the divers as the depth is increased, and hence consumes more gases with increased depth.

Using this method, the diver's body will absorb significant amounts of the nonreactive gas nitrogen that makes up about four-fifths of the air we breathe. There must be nitrogen in this gaseous mix because pure oxygen is a poison to the human body. This pressurized nitrogen content in the bloodstream is only slowly released from the body as the diver rises to the surface and the pressure is reduced. If the pressure is reduced too rapidly, the dissolved nitrogen creates bubbles in the bloodstream and can create the deadly condition called nitrogen narcosis, or the bends. Consequently, SCUBA tanks must have a capacity to both provide the high pressure needed for work at depth and allow the diver to ascend slowly to the surface.

Helium, an inert gas, can be used to partially offset nitrogen in the mix and avoid the problem of the bends. However, in either regard, SCUBA tanks can only contain sufficient gas to allow a diver to work at depth for relatively short periods, counted in minutes rather than hours. The greater the depth, the shorter the duration of the dive.

A rebreather apparatus can extend the usefulness of SCUBA equipment by reprocessing the expelled carbon dioxide to convert it back to oxygen. Using such aqualung technologies, time at depth is increased and problems with decompression are obviated. Even with these technologies, depth records are in the range of 1,000 feet below the surface. Working depths for professional divers are shallower.

A modern atmospheric diving suit (ADS) is currently able to reach a depth of up to 2,300 feet, and allows divers to remain at depth for a significant amount of time. Because the suit is hard shell, and the internal pressure approximates atmospheric pressure at the surface, there is no need for the technique of decompression to ensure the nitrogen saturating the bloodstream from the nitrogen-rich air the diver breathes can be flushed slowly from the body.

For practical purposes, though, any depth beyond 1,000 feet cannot be considered routine. Depths greater than 1,000 feet or more are the purview of robotic Remotely Operated Vehicles, controlled by wire by humans, usually in a ship at the surface, but, theoretically, from anywhere in the world.

The deepwater oil exploration industry has become the world's most adept employers of such high pressure, extreme condition technologies. These highly advanced technologies exceed even those employed in space for construction and repair. The industry works with these highly specialized pieces of equipment, to bolt and unbolt, saw, shear, and even weld, at depth, and to move the levers and rotate the valves that operate complex equipment at the wellhead. The highly technical and rarified nature of this profession caused the United States Federal Government to quickly conclude that it must partner with BP in the deepwater repair effort rather than assume the lead position. An entity could go out and hire its own deepwater experts, but would soon find that it would be hiring the same experts already employed by BP.

In one dramatic moment as the resources of the nation were commanded to repair the well, movie maker and director James Cameron volunteered his services. His offer was based on what he felt was significant expertise he gained from the filming of his epic movie "Titanic." BP respectfully declined his offer because they were, no doubt appropriately confident they already had assembled a state-of-the-technology team. Director Cameron was not amused, and labeled the BP executives "morons."[115]

A well design refresher

The well itself is of a common design. A well hole is drilled as a series of concentric holes, each deeper hole of slightly smaller radius than the previous hole. Once one hole is drilled, a casing, or liner, of steel is placed in the hole and cemented in place to ensure the rock and sand does not later collapse. This cement stabilizes the casing in the wellbore, and, hopefully, seals the outer annulus between the casing and the bore to prevent hydrocarbons from migrating up the outside of the casing.

These casings continue to extend downward, much like the extension of a pocket telescope once used by seafaring explorers. As the bottom of the reservoir is neared, a final casing is cemented and the final hole to the hydrocarbons is drilled. Through the middle of the drill pipe is circulated high pressure drilling mud that lubricates the drill head, circulates away the drilled rock and debris, and fills the void between the drill pipe and wellbore to ensure hydrocarbons do not flow up the annulus between the wellbore and the inside of the series of casings.

This mud must be of sufficient weight to counteract the pressure of the reservoir oil, and must be sufficiently viscous to remain intact and act as a barrier to the oil. Once the reservoir is penetrated, the drill pipe is retracted as more mud is pumped in, and the well maintains a precarious balance between oil and mud.

Finally, a production casing is inserted into the well. This casing is smaller than the steel liner, and creates the path that production oil will migrate to the wellhead once the well is complete. The production casing could be a one-piece contiguous pipe, as BP employed in the case of the Macondo Prospect well, or it could be a two-piece design, with the lower piece hanging from lowest outer casing, and tied back into a production casing that then goes from the intermediate joint to the wellhead. This two-piece, tieback solution has one advantage in that it can be cemented to the final well liner, and hence offers one more barrier to oil that might want to enter the annulus between the well liner and the production casing.

Some have criticized BP for using the one-piece casing, primarily because of its lack of an additional barrier to migrating oil within the annulus. However, subsequent data analysis of the well, both at the static kill and bottom kill stages, to be described later, demonstrated that the problem with the well design was not in oil migration through the annulus, as an alternate tie-back design might have remedied.

In either design, the bottom of the production casing is cemented to the surrounding substrate. This cementing job is the most critical seal of any well, is performed at a place with the highest risk of hydrocarbon contamination, and suffers from the most extreme range of temperature and pressure. From post-accident investigations, it appears most likely that this bottom cementing failed at the Macondo Prospect well and initiated the cascading events that led to the blowout. Indeed, most well blowouts are a result of a failed cement job.

Solutions to the spill

Immediately, following the collapse of the Deepwater Horizon riser pipe, engineers began to devise various plans to stem the flow of oil from the damaged well. Invariably, the initial plans failed. With each failed experiment, the engineers reverted to a plan that many, in retrospect, would utter "why did you not try that first." However, the engineers were initially not trying to stop the well, but were rather attempting to slow the well down so that they could buy some time until the relief well could intercept the runaway well by August. It would be inaccurate to assume that the engineers were either inept or irresponsible. A highly skilled team of BP engineers and federal scientists were trying the best options they could formulate for a circumstance that had never happened before. However, in a retrospective view influenced by what was later found would work, it is easy to declare that the ultimately successful strategy should have been done outright.

The relief wells

If the flow rate of oil is not so significant, or, perhaps cynically, is not so highly visible as was the Deepwater Horizon spill, the most common solution to a blowout in which such safeguards as the blowout preventer have failed is to drill a parallel relief well. This well can either penetrate the oil reservoir to offer a less constricted pathway to the surface and thereby reduce flow to the blown-out well, or can intercept the well so that cement can be injected there to choke it off. This latter goal is called a bottom kill because it typically kills the runaway well near its bottom, just above where it penetrates the reservoir.

The most significant recent attempt at such a bottom kill was at the West Atlas well in the Timor Sea north of Australia, a failure that also implicated Halliburton in a potentially failed well cement job. In that case, it took ten weeks for engineers to drill 11,500 feet, which is, coincidentally, very similar to the depth of substrate the BP team had to drill for its relief well. In the case of the West Atlas relief well, the drilling team actually drilled into a pocket of oil and gas as it neared the runaway well. Fortunately, while the platform had to conduct an emergency evacuation, no fire ensued and no worker was killed. Nonetheless, the experience points to the obvious problem of drilling an almost identical well right next to a failed well. Whatever circumstances caused the original well to fail can also be repeated in the second well.

This pace of drilling at West Atlas was not too dissimilar to that BP attained. It is not unusual to drill two relief wells simultaneously, as Ixtoc I had done more than two decades earlier. However, the Deepwater Horizon response was unique in the number of simultaneous solutions it attempted to employ. Because some solutions at critical stages commended a halt to drilling of relief wells that were within feet of the runaway well, drilling was suspended a number of times. Also, a tropical storm forced a weather suspension for safety reasons. On September 19, 2010, the first relief well intercepted the damaged casing, injected cement into the wellbore and forever sealed the well from below, two months after it had been sealed at its top.

The containment dome

While a number of strategies were developed in the first few weeks of the blowout and many were simultaneously initiated, the first quick fix was in the placement of a specially designed containment dome over the large blowout preventer on the top of the wellhead. It was hoped that this dome would fit over the BOP and seal, at least somewhat so, against the seafloor. If the device worked as planned, the containment dome would then pump escaping oil from a valve at its top so that the pressure of the escaping oil would not thrust the containment dome upward and off of its target.

The pressure of escaping oil in this well was high. Although the containment dome would weigh hundreds of tons, its sheer weight could not counteract the pressure of the oil. Finally, at the time, BP and the government flow assessment team believed the flow from the well was less than a quarter of later and more accurate estimates. Consequently, the success of such a containment dome would depend on its ability to bleed off oil to the surface before the well could cause the dome to rise. This solution could not serve as a permanent solution. Its best hope was that it would temporarily contain spewing oil until a bottom kill, a cement seal injected into the bottom of the well and a relief well could be completed.

Actually, two containment domes were built within weeks of the spill. The engineering team was concerned about the potential success of sealing the dome on the bottom, and the risk of having the dome lift off due to the highly pressurized oil. However, before they would be able to use a dome of any variety, they had to remove some of the nest of riser pipe littering the bottom of the seafloor. ROVs armed with saws and shears, large hydraulically powered scissors that could either cut, or snip shut the pipes, were at work clearing a path.

Unfortunately, as they cut and cleared pipe, they created shorter and shorter paths for the oil to flow out of the well. Estimates of oil loss to the Gulf started off low, and began to increase. An early estimate, attributed to BP but made in conjunction with a federal science team, determined that 5,000 barrels of oil was lost per day. As more pipe was trimmed, this amount crept higher.

With much of the riser pipe removed, a containment dome was lowered onto the BOP and the last vestige of bent-over riser pipe. The containment dome proved ineffective. BP engineers were unwilling to attach a riser pipe to the dome at first because they feared that, without sufficient pumping, there would be lift. In addition, the combination of oil and gas, when it comes in contact with very cold sea water and a drop in pressure, will convert some of the gases and liquids, called hydrates, into a solid with characteristics much like ice. This ice-like solid would both clog up the dome and would cause it to lift.

Ultimately, the attempt to use a containment dome failed, and engineers had to continue to their next plan, the clearing of the last bit of bent riser pipe on top of the BOP and a capping.

The straw

After failed containment cap attempts, engineers sheared off all but the last few feet of the riser standing almost vertically above the BOP. They at first had hoped that a steel saw, attached to one of the ROVs, could cleanly cut the riser so that a tight-fitting cap could slip over it and attach the riser stub to a new riser that could take oil to a waiting ship on the surface. However, the specially designed saw became stuck. Instead, engineers used a large set of shears to cut the saw. The difficulty with shears is that they do not produce an undistorted cut. As a consequence, the cap they fitted over the stub fit less than perfectly.

This solution, at least, was able to eventually divert a share of the oil, even if preparing for the solution may have inadvertently opened the pipe further, and likely allowed oil to escape at a higher rate. Indeed, before this partial capping, the well was still highly pressurized, having little opportunity to deplete yet, and was running out of a stub that offered almost completely unobstructed outflow. The upper end of reasonable estimates of the flow rate by the Flow Rate Technical Group, an assembly of scientists assigned by the Federal Government to calculate flow based on measured pressures and video pictures, at the maximum were as high as 60,000 barrels per day. Images of unimpeded oil flowing into the Gulf of Mexico played on some news channels in real time, constantly.

The "junk" top kill

With a loose-fitting cap siphoning off perhaps 20%–40% of the flow, and partially impeding some of the remaining flow, engineers saw an opportunity to attempt a top kill. This technique involved forcing a large amount of mud, steel balls, cement, and bits of rubber, called "junk", into the top of the blowout preventer. A similar procedure had partially abated the flow to buy time for the team at Ixtoc I to drill its relief wells three decades earlier. However, given the high, perhaps even much higher than expected, outflow of oil, the junk was likely washed out into the Gulf just as quickly as it was pumped in. This attempt, too, failed.

The top cap

The next strategy, given that the well was at least partially contained, was to unbolt the riser beneath the top cap and bolt on a new, tight-fitting riser stub, attached to a new riser tube running to the surface.

Daunting were the technological challenges that would require robots to remove dozens of huge nuts and bolts almost a mile below the ocean's surface, and replace the severed riser with a new riser attachment. All the while, oil was flowing out at a rate of almost 40 barrels of oil, or more than 1,600 gallons of oil per minute. The process went surprisingly smoothly.

The next decision was whether to tap off the oil. At that time, the extent of damage to the wellbore as a consequence of months of high volume and high pressure oil discharge was unknown. If the wellbore was not intact, or if fissures in the soft sandstone or salt had formed, there was a fear that blocking up the flow would cause oil to begin to flow out of myriad places on the ocean floor that would be impossible to remedy. Indeed, blocking off the Santa Barbara spill four decades earlier caused such seepage to continue unabated for more than a year, until internal pressures subsided. And, that field and flow rate was orders of magnitude smaller.

This technique eventually worked by July. Oil stopped flowing into the Gulf of Mexico. With the oil trapped in the production casing and blowout preventer in a static condition, the engineers were finally in the best position to perform a static kill. This static kill involved injecting sufficient cement of optimal density and weight into the wellhead. Because there was no longer any flow of oil, this cement would finally remain where it was injected, and was able to set up. Finally, most engineers were confident that the spilling had stopped.

At this point, it was estimated by the Federal Government's science team that approximately 4.9 million barrels had left the wellhead, and .826 million barrels was recovered above the wellhead. Of the 4.1 million barrels that entered the Gulf, some was dispersed, mixed with the seawater, and is slowly being consumed by oil-eating organisms; some was recovered in oil booms or removed from shoreline; some of the lighter, more volatile molecules of the oil slicks evaporated into the air; and some slicks were burned, while some undispersed heavy oil may still reside on the bottom of the seafloor or in wetland grasses.

Once the flow had stopped, engineers were able to investigate the wellhead to determine if the production casing had lifted, if oil was running up the annulus between the casing and well liner, and if the junk kill had done any damage to the upper well components. They concluded that the well remained sound and the cement cap would hold. The lack of oil in the annulus, especially, gave insight into the type of subsequent kill they would need to do further down the well.

With the well successfully killed, the engineers orchestrated ROVs to remove the failed blowout preventer that was to have saved the well and eleven lives. On September 3, 2010, the blowout preventer was detached and brought to an awaiting ship so that the Department of Justice could perform its investigation of what might have gone wrong and who might be culpable. A replacement blowout preventer was bolted in place.

The bottom kill

Nonetheless, the engineers and scientists were determined to complete the relief well process. In mid-September, 2010, the relief well drilling team, who had been delayed as other, sensitive, processes were completed, finally managed to reach the well. Within a couple of days, the outer annulus had been filled with cement. Finally, cement plugs had been successfully inserted in the annuli and the production casing. Retired Coast Guard Admiral Allen declared the well dead on September 19, 2010.

The aftermath

As noted elsewhere, once the image of spewing oil ended in July, the media quickly lost interest with the story. Also almost immediately, oil slicks dissipated, tar balls washed up onto beaches became rarer, and, surprisingly soon thereafter, fishing grounds were reopened. The image of spewing oil was indelible, however, and tourism and the seafood industry on the Gulf remained depressed, as we begin to discuss next. Quickly, other stories caught the attention of a recession-angry nation as elections neared, and BP left the headlines.

20
The Toll on the Environment

With a final cap in place and the well killed, and with no new oil released into the Gulf, cleanup crews could begin to get ahead of the environmental calamity. Those crews cleaned up beaches, placed, and replaced booms offshore. They had also rinsed grasslands when intervention was more effective, and would induce less damage, than allowing nature to regenerate the marshes. Finally, they could see much more rapid progress.

The Gulf of Mexico imposes some different challenges and opportunities. The ecosystem is rich in diversity. A large human population lived alongside the Gulf and gained its livelihood from the fisheries and tourist attractions on and around the Gulf.

These factors heightened the vulnerability the flora, fauna, and human residents that depended on the Gulf of Mexico. On the other hand, the warm waters of the Gulf, the natural occurrence of organisms that could digest and process oil, and the turbulence of Gulf waters helped accelerate environmental recovery.

Federal scientists released an oil budget that documented the expected status of the spill shortly after the flow was cut off. To generate its estimate, the National Incident Command assembled a team of experts to quantify the amount of oil released from the well and the volumes that were absorbed into the environment. Under the direction Steven Chu, a PhD physicist and Secretary of Energy, and the U.S. Geological Survey (USGS), the team estimated that 4.1 million barrels were released into the Gulf of Mexico. Their inter-agency report stood as the best estimate of the oil spill balances when the report was released on August 4, 2010 (Figure 20.1).[116]

These federal scientific estimates were not without controversy. Some scientists reported in the journal *Science Online* that there may remain large undersea plumes that were not degrading rapidly.[117] However, new

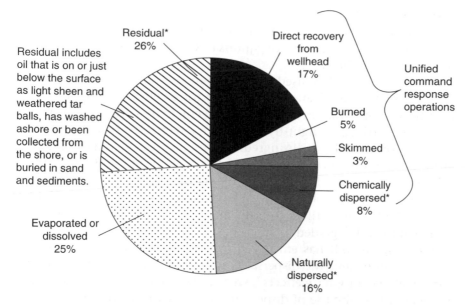

Figure 20.1 Deepwater Horizon Oil budget courtesy of the Flow Rate Technical Group

Note: *Oil in these 3 categories is currently being degraded naturally.

Source: www.noaanews.noaa.gov/stories2010/images/oil-chart.jpg, accessed April 6, 2001.

scientific information was also discovered. While it has been long known that oil will biodegrade, especially if it remains at or just below the ocean surface, Gulf studies discovered a previously unknown oil-consuming bacteria that can function in deeper water, without the oxygen characterizing other biodegradation pathways. In the study reported in the prestigious journal *Science*, the authors suggest that this bacterium at least partially explains the rapid disappearance of oil at the surface.[118]

This degradation was documented to work very quickly. The microbes were able to degrade hydrocarbon alkanes by one half within 1.2 to 6.1 days, an article reviewing this work reports.[119]

The role of dispersants

In the constantly running video images of oil spewing from the ocean floor, a careful viewer could invariably see a wand projecting into the flow as a milky white liquid was released. The reason why this wand, held be the arm of one of the ROVs, was so ubiquitous is that the camera held transfixed to the spewing oil was attached to the ROV tasked with spraying dispersant into the spewing oil. This ROV, operated by the ship holding and distributing dispersants, had the unique responsibility of injecting millions of gallons of COREXIT 9500 into the oil as it leaves the wellhead.

COREXIT is a widely used dispersant in the oil extraction industry. The Materials Safety Data Sheet (MSDS) for COREXIT 9500 deems the chemical, a combination of propylene glycol, to be a nontoxic antifreeze combined with some light hydrocarbon solvents, to have low human and environmental risk. Its ingredients are not considered carcinogens, although no long-term exposure studies have been conducted on the solution.[120] It allows the oil to be broken down into much smaller globules. These smaller globules have a larger surface area per unit weight of oil, and hence allow them to both mix better with water and become attached to and degraded by microbes.

However, no spill has employed dispersants in such high volumes. While many believe the dispersants to be safe and preferable to undispersed oil, the long-term effects have yet to be determined. There is some agreement that the use of dispersants slows the migration of oil to the surface, and may give rise to the plumes reported by some researchers.

COREXIT is also believed to enhance the degradation of oil by allowing the microbes discovered by Hazen, et al., to act at oxygen depleted depths to degrade the hydrocarbons with surprising rapidity. Hazen reported that "These findings also show that psychrophilic oil-degrading microbial populations and their associated microbial communities play a significant role in controlling the ultimate fates and consequences of deep-sea oil plumes in the Gulf of Mexico."[121]

The Gulf ecology

The Gulf of Mexico is a mix of deep-sea and continental shelf waters that is thought to have formed 300 million years ago as a consequence of subsidence. It contains 643 quadrillion gallons of water, for which the oil released represents one part in four billion, which induced then BP CEO to exclaim to his later regret that the spill was tiny in relation to a very big ocean.[122]

With the exception of salmon, the Gulf produces the majority of the major species of fish and shellfish harvested in the United States. It contains about half of the nation's wetland, and is a habitat for three quarters of the nation's migrating waterfowl. More than 400 species of shellfish can be found in the Gulf and it is a habitat for bottlenose dolphins. It also is the habitat for the Loggerhead Turtle and other varieties, and more than 15,000 species of birds, fish, mollusks, crustaceans, fish, and mammals. It also produces a quarter of the nation's natural gas and an eighth of its oil.[123]

As of August 13, 2010, the U.S. Fish and Wildlife Service tabulated 1826 dead birds with visible oil, 17 sea turtles, 4 sea mammals, and no sea reptiles. No dead fish were reported in this tabulation.[124] To place these tragic wildlife deaths in perspective, the Exxon Valdez spill resulted in the immediate loss of more than 35,000 birds and 1,000 sea otters. However, because dead carcasses sink, estimates suggest that the Exxon Valdez spill may have caused the death of 250,000 seabirds, 300 harbor seals, 2,800 sea otters, as many as 22 killer whales, 250 bald eagles, and billions of herring and salmon eggs, according to the Exxon Valdez Oil Spill Trustee Council.[125] To place these numbers in perspective, the Torrey Canyon spill in 1967 killed 15,000 seabirds, while the 1978 spill of the Amoco Cadiz killed 20,000 seabirds and 20,000,000 pounds of oysters.

While wildlife morbidity may lag far behind the more concentrated and confined spill in Prince William Sound of Alaska, some postulate that long-term effects of the spill may still be profound. For instance, contamination of wetlands could adversely affect the spawning grounds for Loggerhead Turtles. In addition, we have yet to discover how the diluted oil and dispersants affect fish and shellfish, or how the microbes that degrade oil will affect the food chain. In addition, the high concentration of the primary natural gas constituent, methane, may have an effect on ocean animals, either directly, or through its effect on oxygen levels in the ocean.

Finally, if, as some postulate, dispersed oil can be deposited on the seafloor, animals such as crabs and oysters, and their larvae, may suffer longer-term consequences. While the Gulf of Mexico appears so far to have remarkable self-cleansing properties, only time will tell if there will be lingering environmental challenges. Obviously, these effects deem further study. We can be sure of one thing, though. The Gulf of Mexico will become the most studied oil spill recovery instance in ecological and biological history.

Ecosystem costs

The costs to the Gulf ecosystem will take some time to quantify, even if indications of more rapid-than-expected recovery are mounting. However, restoration costs may still outstrip the $20 billion BP placed in trust for such remediation. One estimate, made to the President's Commission, by James T.B. Tripp and Brian McPeek place the costs of restoration at $500 million per year for the next 30 years.[126]

Part V
Politics, Courts, and Markets

It would be surprising if a commodity in which the world spends almost $10 billion each day to purchase would not also drive our politics, our litigation, and our financial markets. In this Part, I treat how the Deepwater Horizon oil spill has become politicized, and has likely created a ten-year backlog for the courts to settle. I also document how financial markets may have overreacted to the political and legal hyperbole.

21
The Politics of Oil

Oil is intrinsically political. No other resource commands the interest and passion of oil, the intrigue of wily oilmen, robed sheiks, kings and sultans, roughnecking cowboys, and stiff-suited executives. Jimmy Carter, the president of the United States from 1977 to 1981, won and then lost an election, in large part due to oil. And, the Organization of Oil Exporting Countries (OPEC) redressed global power balances in a seemingly irreversible way when it discarded the colonial model of oil exploitation, asserted its power to control a significant part of the world's oil reserves, and moved the price of a barrel of oil from single digits to triple digits. Oil, and the cars and trucks that consume it, are typically the largest items in countries' balance of trade. Now, many hourly newscasts quote the value of the Dow Jones Industrial stock market average and the price of oil. And, in the year of the Deepwater Horizon spill, China, the world's next oil-thirsty nation, overtook the United States as the world's largest market for automobiles and overtook Japan to became the world's second largest economy.

People have oil on their mind. Military leaders fight wars over oil, and can lose wars over a shortage of oil. And, politicians tap into the concerns of people and the military.

BP's origins sprang from oil. To ensure a steady supply of oil for the Great War, Winston Churchill ordered the nationalization of the Anglo Persian Oil Company, the predecessor of BP. Since then, BP has been intertwined with the strategies of its host nation. BP also remains very popular with Britain's investors and public pension plans, especially since it was privatized over the period 1979–1987.

While BP may be viewed in more nationalistic terms within the British economy, part of this status no doubt comes from a pride of self-sufficiency and appreciation for the jobs and contribution BP makes to the British economy. People know not to bite the hand that feeds them. And, as one of the world's largest companies, at least just before the spill, and the largest company in Great Britain, BP inevitably has the ear of its government.

On the other hand, the American public has a much more uncomfortable relationship with big oil. From early attempts by Rockefeller-owned Standard Oil to monopolize all oil, from the wellhead to the gasoline station, the public has witnessed oil companies as monopolistic pariahs that would gobble up mom-and-pop operations so that they could control the price of a commodity everybody needs. The antics of Standard Oil resulted in Congressional action to break up such monopolies and trusts. From the roots of Standard Oil came a namesake, Esso (for S.O.), and Exxon. And, when Exxon Valdez created such indelible images of dead birds and seals in pristine waters in 1989, followed by Exxon's shirking and one of its most profitable years ever, the American public began to mistrust oil, even as it had to begrudgingly fill up its gas tank every week.

In some sense, the American public's response to the BP spill must be viewed in this negative light of Big Oil. Compounding the American reticence was the pain a nation was experiencing because of the failure and malfeasance of Big Finance, a folly that had plunged the United States, and much of the developed world, into a recession that was the worst since the Great Depression.

In such troubled economic times, it is typical that nations will circle their wagons. Clarion calls along nationalistic themes of protectionism and mercantilism inevitably emerge and demonize all things "foreign." In the United States, migrant workers who have been travelling from Mexico to help harvest U.S. crops and take jobs a fully employed labor force refuses to accept all of a sudden became highly resented by a significant part of the public. Indeed, the three most followed stories in 2010 were the BP spill, calls for a get tough policy with "illegal immigrants," and the plight of the economy. And, a new political force emerged. The "Tea Party" that became a U.S. political fixture in 2010 invoked the sentiment of early American revolutionaries that took the law into its own hands, seized shipments of tea from British ships in the Boston harbor, and threw them overboard in an act of defiance against their British rulers. America was in a mistrustful mood.

This confluence of forces could not have been more unfortunate for BP, or, as it was popular to iterate at the time, British Petroleum, a term that was much more laden to invoke foreign and oil. This term was a double whammy of negative connotations for the American public, as a constantly looping video of the flow of oil into a Gulf of Mexico was playing in the corner of every newscast screen to drive home the point.

Adding to this explosive mix was a collective guilt about how the region was neglected in 2005 in the aftermath of Hurricane Katrina. Residents of the Gulf have often felt the neglect and, perhaps, some incipient racism. This public shame was in full view over the lack of preparation, over many years, by the federal government's Army Corp of Engineers in their construction and maintenance of the levies that collapsed during Katrina. The Federal Government could not be perceived, once again, as neglecting the Gulf of Mexico in a time of need. Black tar balls rolling up on pristine white beaches was just the symbol the Federal Government had to avoid.

Fortunately, there was a party that accepted responsibility for the spill and its consequences early on. BP's acceptance of financial responsibility was, no doubt, calculated to get ahead of the tragedy before it really spun out of control. It was not an admission of culpability over the spill, though. Indeed, BP had partners in the field and had contracted with two other deep pocket corporations in critical aspects that led to the spill. Rather, BP's acceptance of financial responsibility was to lay to rest the speculation that the spill would not be fully remediated, and the economic consequences fully compensated over and above the $75 million statutory liability.

BP was, in essence, saying the public could have its ounce of blood, but let's not take a pound of flesh. And, yet, an angry and impassioned public sought a pound of flesh and then some. All the while, a president weakened in the polls by an ever-worsening economy, no doubt calculated that a strong and tough approach to BP was a political imperative.

The first days of the spill

It is apparent that no entity truly understood the extent and potential of the spill in the first few days following the explosion on April 20 and the riser collapse on April 22. While the Gulf of Mexico major oil companies each maintained a strikingly similar spill response plan, no group truly anticipated the simultaneous series of breaches that could cause such a massive spill.

The Minerals Management Service had required a worst case estimate of the potential size of a spill should the blowout preventer be opened completely to the ocean. This reported potential spill rate was never seriously contemplated. BP had reported that the maximum possible spill rate could be 162,000 barrels per day for this well; these internal and permitting estimates bore no relation to a typical spill under more manageable conditions. BP would refine this estimate to take into account actual riser conditions. The revision, reported by BP to Congress in mid-May, placed the spill at upwards of 60,000 barrels per day, a figure that was remarkably in line with more refined third party estimates provided to government officials much later.

Indeed, no deep water spill had ever come close to that level. Consequently, while Retired Admiral Thad Allen reported to the public in early May that the worst case scenario from the permitting process could result in such a spill rate, the earliest estimates of the flow rate were significantly smaller.

The issue of the state of the spill rate was highly political. Estimates were made in an atmosphere of considerable uncertainty. In October 2010, the National Oil Spill Commission released a report that documented some of the debate that occurred early in the spill.[127]

The report documented that the first somewhat scientific estimate of the spill was reported by a National Atmospheric and Oceanic Administration scientist to be 5,000 barrels per day, based on visual data from the most significant leak in the riser pipe. That April 26 estimate was modified upward the next day by John Amos, the founder of SkyTruth.org, an organization that uses aerial and satellite data to estimate spills. The revised estimate based on the size of the emerging slick upped the spill rate to between 5,000 and 20,000 barrels per day. Then, on May 1, Florida State University oceanographer Dr. Ian MacDonald used an accepted spill estimation protocol, called the Bonn Convention, to place the daily spill rate at 26,500 barrels. However, it was noted that these estimates were somewhat uncertain because aerial data does not well measure the thickness of the slick.

Another set of estimates by mid-May, based on analysis of the video pictures of the spill, and adjusted for the fact that the flow was a mixture of both oil and natural gas, placed the flow at between 10,000 and 50,000 barrels per day. On August 2, 2010, the Flow Rate Technical Group provided the most accepted spill rate figure, of between 52,700 and 62,200 barrels per day, plus or minus 10%. It is apparent that BP's estimate of 60,000 barrels in the first weeks of the spill proved remarkably accurate.

These various estimates were relevant for two reasons. First, government officials stood accused much later of withholding potential spill rates from the public for political purposes. The National Oil Spill Response Commission had reported that operational level officials were told by higher authorities, presumably from the White House, not to report the higher speculative flow figures. The commission postulated that such a withholding, if it existed, would only have made more cynical an already skeptical public.

The greater concern, though, was whether a lower publicized spill rate would have reduced the sense of urgency in the spill response.

Politics of the early response

Certainly, all estimates of potential flow rates would inevitably be discussed between BP and the Unified Command, the team of government officials and scientists assembled to coordinate the spill response. BP had almost immediately accepted financial responsibility for the spill, and well understood that fines under the Clean Water Act were tallying at a rate of more than $1,000 per barrel spilled. With fines falling conceivably in the range of $50 million to $200 million per day, BP and the Unified Command well understood the stakes.

They also understood that BP's liability would not end with any certainty until a bottom kill was successfully completed. Consequently, the primary strategy in these and similar spills is to immediately begin drilling at least one relief well. BP chose to drill two such wells. The first to reach a depth sufficient for a bottom kill could be diverted to intercept and plug the bottom of the well. The second well could act as insurance should something go wrong, and could proceed to tap into the reservoir to relieve reservoir pressure if necessary. Teams of experts from Boots and Coots that specialize in drilling such emergency relief wells immediately got to work.

All other measures are designed only to slow down the flow until the final bottom kill could be made. Such a slowdown of flow reduces both the total amount of oil released, and BP's financial liability under the Clean Water Act. There is no doubt that actions were confused and less than perfectly coordinated in the early hours and days of the spill. However, there is also no doubt that both BP and the federal government wanted the spill abatement to proceed at all speed, even if their motivations differed.

The quandary, though was that the Federal Government could not plug the leak and clean up the shoreline without BP's assistance and expertise, and did not want to assume the role of the party responsible for the pace of the cleanup or capping of the well. They had to form an uneasy alliance, which BP likely appreciated and valued, but which challenged the political optics of the time. The Federal Government viewed BP as a "frienemy," a term popular among teens at the time for a friend who is also an enemy. The Federal authorities needed to demonstrate they were in control, but could ill-afford weakening the company so much that it could not live up to the financial responsibilities it accepted.

Even further complicating an already murky and logically inconsistent political landscape was a growing acceptance of oil's role in global warming, an effort to regulate greenhouse gases as a pollutant under the regulatory authority of the U.S. Environmental Protection Agency, and the introduction of an Obama-sponsored energy reform bill that had been stalled in Congress. This bill had made some concessions to the domestic energy industry as a price of passage. However, after April of 2010, few congressmen felt comfortable championing such a bill. Instead, Congress devoted their time to what they often do in times of populist anger. They held hearings, perhaps prematurely, to ostensibly investigate the cause of a spill that had not yet been stopped, to force executives from Big Oil to come testify in front of the cameras and the American people.

And, the Obama administration sponsored a moratorium on drilling. In doing so, it ratcheted up the stakes still further, and increased the potential economic liability for workers that would become displaced as a consequence of the moratorium imposed in response to the spill.

Obama was actually on loose footing in his moratorium on offshore well exploration. He had campaigned on American energy self-sufficiency and had advocated for greater domestic oil production as an integral part of U.S. energy independence. Accordingly, early in his presidency, and just before the Deepwater Horizon spill, Obama had advocated the most extensive expansion of offshore drilling in recent presidential history.

Obama had even used this expansion in offshore drilling as the carrot he thought he would need to pass a more expansive energy bill that, for the first time, would include carbon taxes to reduce greenhouse gas production.

This expansion of drilling in offshore waters is risky by its very nature. Waters for new wells are deeper and more remote. And, new fields that had previously been off limits are also much larger.

None of these policy considerations at all negate the anger that the American public vented over a spill that was sullying the Gulf. Obviously, BP, like all oil companies, anywhere, was woefully ill-prepared for an environmental disaster of this scope. Certainly, after this experience, oil companies that fail to be more prepared in the future will do so at great peril.

In a political system in which Congress often serves in populist ways, much of the politics was inevitable, if opportunistic. And, some of the pressure brought to bear may have accelerated spill abatement and economic claims adjustment, marginally. However, once BP had accepted financial responsibility, it well understood that a quick end to the spill, and a rapid return of the region to normality was in its best interest.

To the British people, the political folly appeared somewhat contrived, opportunistic, and a gesture of "piling on." Outside of the constant barrage of American media that was thoroughly enjoying the drama, the ratings, and the advertising dollars, a broader view of the entire event, from a more global scale, was inevitable. And, after all, when politicians were punishing BP, they were also damaging the British people.

Political consequences

Obama almost immediately placed a moratorium on new offshore well explorations in the United States, to run until November 30, 2010. The moratorium was successfully challenged in federal court in Louisiana, the Gulf of Mexico state that was fearful a moratorium would do even more damage to their biggest economic sector. However, a replacement moratorium stuck.

Obama also insisted that BP forego dividends to its shareholders while oil remained on the surface of the Gulf of Mexico. From a financial standpoint, such a suspension of dividends did not at all affect the value of the company. BP would merely be retaining earnings rather than distributing them. However, because many holders of BP stock do so to maintain a modest flow of dividend payments each year, many pensioners and pension funds were hurt by this price extracted by the Obama administration. The stock fell somewhat in reflection of the resulting movement of preferences away from the stock. On the other hand, when Bob Dudley, the CEO who replaced Tony Hayward, effective October 1, 2010, announced that dividends would likely resume in 2011, the stock began to rebound slightly.

Obama also demanded that BP put into trust $20 billion over four years to pay for economic claims. The negotiation of that fund was intense. Obviously, while BP easily had the means, from cash flow alone, to fund its commitment, it did not want to make the asset sales necessary to fund the commitment up front. It also likely extracted some understanding that the Obama administration would show restraint in its pursuit of BP in criminal court because any resulting fines levied would not be tax deductible by BP. Two decades earlier, this strategy was successfully employed to allow Exxon to escape criminal liability.

These ongoing discussions of liability and culpability, sufficient gestures of culpability and blame, but not too much, even ensnarled the British government.

At the time of the explosion in April of 2010, British Prime Minister Gordon Brown was already immersed in an election bid for which he would ultimately fail. There was little yet known of the extent of the BP spill and not yet much that a British prime minister could do.

Nor was Brown necessarily willing to use a lot of his scarce political capital to engage in the American political morass surrounding the spill. Even if he were willing to expend scarce political capital, the relationship he maintained with President Obama did not appear to be warm, even if it may have been mutually respectful. Brown was a mercurial leader, while his American counterpart was known as "no drama Obama." Brown lost the election and the prime ministership.

Upon his appointment as prime minister on May 11, 2010, over a coalition government, Mr. David Cameron had his pan-Atlantic work cut out for him. When he made a two-day visit to Washington on July 20–21, Prime Minister Cameron's private discussion with President Obama likely impressed on the President that orchestrating an anti-BP campaign is not in America's interest. Obama's legacy is about the long term, and short-term grandstanding should be left to lesser politicians, if anyone at all. In addition, while BP is owned by many British residents and pensions, it is also 40% owned by U.S. residents. In effect, damaging BP is damaging pension plans and citizens across the country.

Nor would the political message of corporate bashing go over well for a president already accused of demonizing business. After all, a healthy relationship with businesses that want to be good, or better, corporate citizens is necessary for Obama's economic agenda.

Complicating the discussion was the revelation that BP may have intervened to recommend the release to Libya of one of the terrorists jailed after the Lockerbie airplane explosion. While the Scottish authorities claimed that the release of the Lockerbie bomber, Abdel Baset al-Megrahi, was made on humanitarian grounds, BP was accused of lobbying to promote its interest in expanding its commercial relationship with Libya's leader Muammar Abu Minyar al-Gaddafi. This potential conflict was being raised by some U.S. senators just before the Cameron visit. And, while the Scottish authorities that valued their autonomy vis-à-vis their English counterparts may have wondered how the incident involved Cameron, or perhaps even BP, the issue had to complicate the discussion.

Obviously, newly elected Prime Minister Cameron's mettle was tested at those meetings. Following the meeting with Obama, the tone from the White House did seem to soften, with a greater emphasis on cooperation and shared concerns.

A presidential commission

In an effort to remove some of the politics from an obviously highly charged and very public calamity, President Obama appointed a commission to study the events leading up to and following the spill. The goal was to ensure that no such spill can despoil U.S. or other offshore waters again. This blue ribbon committee, the National Commission on the BP Deepwater Horizon Oil Spill and Offshore Drilling, made up of experts and executives, politicians and policymakers, made an observation toward the completion of their deliberations. On November 8, 2010, chief counsel Fred H. Bartlit, a noted trial lawyer hired to advise the commission, stated:

> To date we have not seen a single instance where a human being made a conscious decision to favor dollars over safety...A lot has been said about this, but we have not found a situation where we can say a man had a choice between safety and dollars and put his money on dollars.[128]

Clearly, the committee was telegraphing to the world that they cannot expect simplistic solutions to complicated problems. Indeed, to imagine that failures in complex systems can be reduced to the creation of some new, simple, and static regulatory provisions would be dangerous. Instead, the commission would go on to demand a much more sophisticated and nuanced system of risk management, with operators expected to take on a much more significant role.

However, the final phase, judgment of BP in courts of law, will make the final, if prolonged statement. It is this legal quagmire we turn to next.

22
A Complicated Legal Quagmire

The final word on the Deepwater Horizon spill will be spoken in court, or in a number of courts. The legal issues are intricate. Further complicating proceedings will be a desire for plaintiffs, from the Federal to state governments, commercial fishers and tourist operators, and the families of those lost at sea when the platform exploded and caught fire, to have their day in court at a venue of their choosing. Ideally, each plaintiff, or small group of plaintiffs, would like home court advantage. Alternately, BP and other plaintiffs will try to consolidate proceedings into larger classes, and adjudicate them in a courtroom in a city that appreciates Big Oil. Meanwhile, BP will make every effort to pull its partners into court with it, or will try to sue its partners on the side.

Congress will also attempt to pass laws retroactively that will extract the intended political toll from BP. The Obama administration sponsored a moratorium on drilling. In doing so, it ratcheted up the stakes still further, and increased the potential economic liability for workers that would become displaced as a consequence of the moratorium in response to the spill. It is these issues we will unravel next.

Criminal court

Let us begin with the most serious, and potentially the most expensive, proceeding. The U.S. Department of Justice (DoJ) must decide if it wishes to pursue criminal charges of gross negligence against BP in the deaths of the workers. The challenges of a criminal sanction are twofold. First, the government would have to prove that BP acted with wanton disregard for human life. In criminal cases, the state of mind of the actor, in this case, BP through its agents, must be determined to satisfy "mens rea," or a guilty mind. In essence, the DoJ must establish intent on the part of BP.

It must also meet this burden "beyond a reasonable doubt." Given the circumstances of the case, the shared responsibilities of BP, Transocean, and Halliburton, and the multiple barriers that must have been breached for the accident and deaths to occur, criminal culpability is highly unlikely.

However, if the DoJ were able to prove criminal culpability, it would recommend to the court a level of fines that would act as a sufficient deterrent for those contemplating similar decisions in the future. A corporation cannot be put in jail, and the jailing of executives of corporations is highly improbable. Hence, the DoJ would likely petition for fines in relation to the environmental and economic damages suffered. According to the legal economic theory of optimal deterrence, such fines should be proportional to the damages and the probability of detecting such future transgressions. Because it is highly unlikely that a major spill could go unreported, a deterrent does not need to be escalated when the probability of detection of the crime is very high. Accordingly, the level of fines would likely be proportional to the level of economic damages imposed on innocent parties.

Indeed, every oil company has individually performed each practice for which BP stands accused and had individual failures of most all of the same critical barriers. It would be difficult to convict BP based on these individual elements without also blaming others for past or future failures that are not unusual in the industry. One could differentiate between BP's alleged transgressions and those of its competitors based on the totality of errors. However, any operator that suffers a similar blowout because of a totality of failures is then equally culpable.

If the DoJ cannot establish criminal culpability, BP will benefit in two ways. First, the subsequent civil sanctions could not rely on a guilty conviction invoking gross negligence to establish the same conclusion in civil court. The second way in which BP would benefit is that sanctions imposed in civil court are tax deductible. One cannot deduct on one's taxes the penalty society imposes on an individual deemed to have committed a misdemeanor or felony against the people. However, civil proceedings are regarded as disputes between persons within society, even if the DoJ tries the case on behalf of a group seeking redress. Civil penalties are then transfers of income between individuals, and are considered a tax-deductible cost of doing business.

Strict liability

It is common in complex and inherently risky enterprises to deem an individual or a group the "responsible party" regardless of contributory fault. The reason for such a declaration by law is to make clear to all who is the coordinating party in the best position to ensure a safe operation. For instance, those that purchase, store, and use explosives are considered responsible for those explosives even if others are negligent in their use or even if children break into the storage building and harm themselves. This strict liability is necessary to ensure due care and avoid blame shifting and shirking.

Such strict liability does not mean that others cannot ultimately share in damages. However, by declaring a responsible party, courts induce greater care than would otherwise likely occur, and create a clear roadmap should these courts subsequently be involved in claims following an accident. In the case of oil exploration, the owner of the well is the responsible party for the purposes of subsequent discovery and trial.

Civil liability

Maritime law requires the responsible party of any spill to cover the cleanup and oil recovery resulting from the spill. This responsibility is not in dispute, even though it is likely that BP will have to either settle or sue its partners Anadarko and Mitsui to ensure they pay their share of cleanup. In addition, BP may attempt to recover some of its costs from Transocean, Halliburton, and Cameron International if they are deemed to share in negligence.

The responsible party or parties are also liable for migratory birds affected by the spill. These fines can be up to $15,000 per violation. However, given the relatively modest number of birds found to have been oiled and which died, these fines may be comparatively modest as well. By August 13, 2010, 1,826 birds were found dead, with oil as a responsible or contributing factor.[129]

In addition, the Endangered Species Act, which may apply to some of the 17 sea turtles found dead and oiled, can result in a fine of $25,000 per violation.[130]

Fines for violations of the Clean Water Act will certainly be levied. It is illegal to discharge pollutants into U.S. waters. Any discharge can result in a fine beginning at $1,100 per barrel spilled. If gross negligence can be proven, these fines can run as high as $4,300 per barrel. With an estimated unrecovered volume of oil spilled of approximately 4.1 million barrels, these fines can range from $4.51 billion to $17.6 billion.

The various companies can certainly afford these fines. BP held $6.8 billion in cash and its equivalents early in 2010, while Transocean held $1.6 billion. Halliburton had $1.38 billion and Cameron International

$1.6 billion. In addition, each of these companies, and especially BP, generate sufficient cash flow that the inevitable survival of any of the firms is not in question, given the likely process and possible settlements that can be pursued.

Typically, responsible parties in such events will negotiate with the government in lieu of a trial. We saw in 1989 that Exxon conducted extensive negotiations with the federal government both to limit its liability and to ensure it avoided criminal sanctions that will not be tax deductible. However, these negotiations would likely take years, especially because all the economic damages and environmental consequences will not be known with some certainty for a while.

Potential liability caps

Initial speculation suggested that BP may be protected by a $75 million cap on economic damages, imputed as part of the Oil Pollution Act of 1990 (OPA) that Congress passed in the wake of the Exxon Valdez spill. This is a cap only in economic damages, not the cleanup cost liability imposed under maritime law. BP had been billed $581 million by the Federal Government for cleanup-related expenses as of October 13, 2010 under provisions of the Oil Pollution Act of 1990.[131] They have also committed to spend almost $400 million in a bid to help redevelop wetlands that have been degrading for decades, but which have also been oiled from the spill.

BP could exercise this cap only if it, or its contractors, did not act in gross negligence or violate any federal regulation. Because it is highly likely that some regulation, or health and safety law, was violated, this cap will not likely hold. It is possible that BP may be able to countersue one or more of its contractors for any payments it makes in excess of the $75 million cap if it is found BP did not violate any of these laws or regulations even if its contractors did. However, on October 19, 2010, BP waived its $75 million liability cap under the OPA.[132]

BP's acceptance of responsibility, its promise to return the Gulf to its pre-spill state, or its statement that it would honor all legitimate claims do not necessarily constitute a waiver of its right to the statutory cap on economic damages. BP probably secured through negotiations with the Obama administration an understanding that will be honored by both BP and the federal authorities. These negotiations will likely result in BP honoring its public commitment to cover all legitimate economic damages. Were they to do otherwise, the cost to the company in public relations and goodwill would outswamp any savings through the statutory liability cap.

Actually, the OPA has made it easier to recover economic damages post Exxon Valdez. Before the act, maritime law required the responsible party to only pay for direct economic damages for property damaged by the spill. A tourism-related industry damaged by the loss of business but who did not own beachfront land damaged by the spill would not have been eligible for compensation before the OPA.

The Clean Water Act

Under the Clean Water Act, citizen groups can file a private enforcement suit. A consortium of environmental groups filed such a suit in Louisiana courts demanding damages of $19 billion. These interested parties are able to stake claim against the responsible party to provide the deterrent sufficient to prevent such violations, and to provide an incentive for private parties to monitor other private parties. These suits will likely be rolled up into other suits under consolidation (discussed later).

Proximity suits

Perhaps the most novel and threatening liability facing BP is in proximity claims. A proximity claim is one that is made by someone who asserts economic damages because of the decisions of others who acted, even irrationally, to fears of oiled beaches. For instance, while little oil impinged on the beaches in Florida, tens of thousands of claims could flow from Floridian tourism-related businesses that claim they lost tourist dollars because of bad publicity arising over the spill.

Proving a proximity claim is an uncertain process in this case. The spill occurred when the United States was immersed in a deepening recession, as evidenced by a rapidly eroding index of consumer sentiment in the months following the spill (see Figure 22.1).[133]

To claim that the loss of tourism business was due to the spill and not an economy with eroding consumer confidence and travel is difficult.

A business could instead claim that their business has dropped off more rapidly than a similar business in an unaffected area. However, tourists may be diverting to the unaffected area, thereby raising its tourism spending dramatically relative to the affected area. Consequently, such a comparison exaggerates the true effect of reduced tourism spending relative to a reasonable baseline.

While the legal theory around proximity claims has yet to mature, it seems unlikely that the courts could entertain such damage claims arising from the irrational response to such a spill. The precedent that

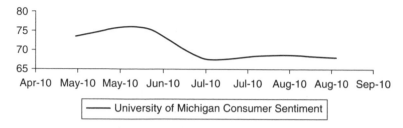

Figure 22.1 The Consumer Sentiment Index in the U.S. in the months following the spill

would be established by the acceptance of such proximate damages would dramatically escalate liability for any similarly situated plaintiff in the future. Nonetheless, until the status of these proximate claimants is determined, a dark cloud may remain over BP and its codefendants.

An unconstitutional bill of attainder?

The U.S. Constitution, Article 1, Section 9 states:

No bill of attainder or ex post facto Law shall be passed.

The principle of this protection is that laws are determined to be followed. Instead, to impose a new law retroactively is to sacrifice the principle that laws should be guiding behavior, not extracting revenge. The civil system works on this premise of equal protection under the law.

In rare circumstances, the courts have allowed Congress to pass a bill retroactively designed to affect a single entity after the fact. Congress would need to demonstrate a compelling state interest to do so, assuming that BP wants to go through the public and media ordeal of challenging the U.S. government in a U.S. court. The corporate calculus would likely prevent BP from rightfully asserting itself to avoid a liability in excess of that provided by OPA 1990.

Mitigation of damages

Another avenue that would allow BP to argue for reduced financial responsibility is in the principle of mitigation of damages. The principle creates the expectation that a compensated victim will not take actions that will escalate the ultimate level of damages the liable entity must absorb. For instance, if you receive damaged goods, and you are entitled for repair or compensation from the shipper, you must take care to ensure the goods are not damaged any further in your custody.

In this case, by imposing an exploration moratorium in the Gulf of Mexico, the federal government ratcheted up the economic damages BP would have to pay to compensate oil workers that would otherwise have been employed. BP was, in essence, on the hook for damages well beyond its control or anticipation.

Again, however, it seems unlikely that BP would challenge the federal government over this issue. In this case, as opposed to the difference in the $20 billion economic damages trust fund compared to the OPA limit, the payments to displaced oil workers as a consequence of the moratorium may tally only in the tens, or perhaps hundreds, of millions of dollars.

The Obama administration announced on October 12, 2010 that it would lift the drilling moratorium, ahead of schedule, for companies that can verify that they have processes in place to prevent a spill from occurring in sensitive offshore waters.

Third party adjudication by a special master

A clear definition of the standard of evidence of economic damage does not exist. Indeed, it is likely that this spill will provide precedents for future spills. To extricate itself from the obvious conflict of adjudicating many such claims, BP agreed to the appointment of a neutral special master, Kenneth Feinberg, to administer any economic claims. Mr. Feinberg was given the freedom to set criteria and evidentiary requirements and would secure a waiver from those that settle which would prevent them to also sue for damages later. Mr. Fienberg set a standard by which all similarly positioned claimants will be treated. His standard would likely not depart significantly from those that a court would impose, but would offer those claims much more quickly and without the need for claimants to hire lawyers and pay legal fees.

One challenge Mr. Fienberg faced early on, and perhaps BP had already discovered, is that many claimants had worked on a cash basis, without receipts or federal income tax returns that accurately measure income. Mr. Fienberg was willing to accept claims of income nonetheless, but was reported to still require some quantifiable indication or some respected third party affidavit of income.

One group that will likely be able to substantiate lost income or increased expenses will be states and municipalities that lost revenue because of lost tourism, foregone income or sales taxes, incurred expenses from first responders, or services offered in conjunction with the cleanup effort.

These various groups can either settle through the special master, or must await the completion of administrative channels before they enter court proceedings. As such, a group interested in timely resolution would best look to the administrative framework, rather than litigation in court. After all, litigation for some took more than two decades following the Exxon Valdez spill. Some claimants have still not received resolution from the courts.

Cap on Transocean's liability

Perhaps the most interesting invocation of a liability cap was filed by Transocean shortly after the accident. In a preemptory move in federal court in Houston and in anticipation of suits for personal injury and wrongful death from its employees that went missing and are presumed dead following the accident, Transocean invoked a maritime law dating back to the pre-civil war era in the 1800s U.S. This law limited liability to the value of the vessel and its freight. Obviously, a vessel now on the bottom of the Gulf, at a depth of more than 5.000 feet has no value. However, at the time of the explosion, Transocean noted that BP owed them $27 million. This imputed recoverable value of freight then acted as Transocean's liability shield, according to their claim. It remains to be seen whether the court accepts this asserted liability shield.

This cap is only on economic damages, in this case for injury and wrongful death. The cap does not trump OPA's imposition of liability for Transocean's share of the oil spill. Nor does this cap extend to BP. It applies only to vessel owners.

Shareholder suits

Almost immediately after the spill, lawsuits began to be filed by law firms attempting to create a class of plaintiffs. For instance, Zwerling, Schachter & Zwerling, LLP, filed a class action suit in the U.S. District Court for the Eastern District of Louisiana. The suit cited a history of safety lapses and cost cutting measures that would have damaged shareholders who may have owned stock between February 27, 2008 and May 12, 2010. The suit contends that BP representations to the Securities and Exchange Commission acted as a basis for plaintiffs' purchase of stock, and that misrepresentations by BP caused economic damage to these plaintiffs.

At the time of the suit, the law firm was still seeking a lead plaintiff. Such suits are somewhat parasitic in that they seek only a transfer of wealth from all shareholders to the subset in the class. The industry of such suits is in the fees they generate for the law firm that champions the suit on a contingency basis or as a special award by the court. They may also try to force the defendants to also cover legal fees, in addition to the sum they must move from the value of stock of their current shareholders to the class action group.

Often, these suits are designed by firms that specialize in class action securities suits. An inspection of the website of the firm, Zwerling, Schachter & Zwerling, for instance, will list numerous similar suits and claims, in various states, with various companies.[134]

A suit against fireboat first responders

Finally, a suit was filed on behalf of fisherman in the U.S. District Court, Eastern District of Louisiana in New Orleans. The legal theory asserts that the flooding of water by fireboats destabilized the rig and caused it to list and then sink. The suit alleges that up to 8 fireboats were each pumping between 40 and 200 tons of water per minute onto a platform that was not designed to support that weight in water. By inundating the upper compartments of the platform with hundreds of thousands of tons of water, rather than with materials designed to extinguish oil fires, and by failing to use the fireboat systems to hold the platform in place, it has been argued that their actions caused the platform to sink and the riser pipe to sever.

Range of economic damages

Negotiations between BP and the Obama administration resulted in the formation of a $20 billion trust fund to compensate those who experience verifiable economic damages. President Obama appointed the special master, Kenneth Fienberg, to administer the claims process. While Mr. Feinberg did not begin until August, by September 25, 2010, he had paid out $379,259,484 in claims to 27,998 claimants.[135] At the same time, though, the administration has criticized the process as proceeding at an unacceptable pace.

In a public report on September 23, 2010, BP stated that it had paid out $1,771,002,869 in economic claims to individuals and businesses, response and removal grants, damages payments, and to governments.[136] It is estimated that BP will have paid out over $2 billion in payments for compensation and economic damages by October of 2010.

While BP devoted $20 billion to compensation for economic damages, and approximately 10% of that sum has been paid out in the first six months since the spill, accurate estimates of the total cost of economic damages will likely be unavailable until late in 2011.

Consolidation of suits

Meanwhile, hundreds of pending federal lawsuits were consolidated and will be considered in the Eastern District of Louisiana in New Orleans.[137] While BP would surely have preferred consolidation of suits in Houston, Texas, where its U.S. headquarters is located and where juries appreciate Big Oil, New Orleans will be both sensitive to the losses of its residents, but almost equally appreciative of the oil industry. Indeed, the moratorium against new offshore oil exploration was overturned in New Orleans by a judge, Martin L.C. Feldman, who, like many financially secure individuals in that region, owned stock in a Big Oil company.

Consolidation of suits, in front of Judge Carl J. Barbier, began in New Orleans in September of 2010. By October, Barbier's court began the process that would name 12–15 lawyers to represent plaintiffs of all claims, on behalf of fisherman, families of those lost on the Deepwater Horizon rig, tourism and visitor industry operators, property owners, and others. Hundreds of lawyers vie to be named to this exclusive team so that they can earn part of the 15% court award for legal fees and contingency fees typically of 30% that go to lawyers from the eventual court-ordered awards. With upwards of tens of billions of dollars at stake, these legal fees could tally eleven digit figures that exceed $10 billion.

Once appointed, this exclusive team of some of the top lawyers in the country will be responsible for the legal strategies and decisions on behalf of all those legally harmed by the spill. This strategic and legal consolidation has many concurrent functions. It avoids the cost and inefficiency of duplicative discovery, the process of obtaining evidence and statements that many of the suits would otherwise have to collect individually. It also provides for greater convenience of the court and witnesses, and prevents contradictory rulings by various courts that would otherwise be spread over half a dozen jurisdictions.

This consolidation would also bring in Transocean, Halliburton Energy Services, Inc., and Cameron International as codefendants in as many as eighty primary suits and a couple of hundred suits in total. These latter suits can name themselves interested parties in the primary suits.

Range of estimates of the extent of the spill

The most significant costs to BP will be related to the fine it negotiates or settles with the U.S. federal government. However, while BP did not publish estimates of the spill rate, it did work with the National Oceanic and Atmospheric Administration, the Bureau of Ocean Energy Management, Regulation and Enforcement (formerly the Minerals Management Service, or MMS), the Flow Rate Technical Group (FRTG), a group of scientists appointed by the Secretary of Energy, and others who provided estimates of oil flow.

These estimates varied widely, for a number of reasons. First, the amount of oil discharged each day changed. As the pipe was cut and hence was less constricted, flow would increase. As the reservoir was depleted, its pressure would fall slightly, and the flow rate would decrease. The flow that BP had estimated before the spill as a condition of their drill permit application, based on a wide open and unrestricted blowout preventer, would be the highest possible potential flow and exceeded the actual flow at any tim during the spill.

Estimates also varied depending on the method used to calculate flow. Some estimates were based on the relationship between the size of slicks and thickness of oil slicks and the flow. Other calculations attempted to model the temperature, pressure, viscosity, and velocity of oil seen spewing out of the riser pipe of known diameter.

BP will negotiate with the science team and federal government to produce a final calculation of leaked oil. This estimate would be a sum of daily rates at various stages of the leak. The best estimate of the FRTG of the total amount of oil leaked and unrecovered at the well site is 4.9 million barrels. Later, the first of a series of peer-reviewed articles

Table 22.1   Various public estimates of flow rates over the timespan of the spill

Source of estimate	Date	Barrels per day
BP estimate of maximum spill rate with no blowout preventer	Permitting stage	162,000
United States Coast Guard	April 24, 2010	1,000
Official estimates	April 29, 2010	1,000–5,000
Official estimates	May 27, 2010	12,000–19,000
Official estimates	June 10, 2010	25,000–30,000
Flow Rate Technical Group	June 19, 2010	35,000–60,000
Official estimates	August 2, 2010	57,500–62,000

appeared that confirmed the range of estimates by the FRTG. Using a technique called optical plume velocimetry, Professor Timothy Crone of Columbia University, and his colleague Maya Tolstoy, estimated the amount of oil that escaped into the Gulf to be 4.4 million barrels. Their article appeared in the journal *Science*.[138]

The range of daily estimates for the leak at various dates is given in Table 22.1.

23
The Market Response

In the heyday of rising oil prices, and just four years before the Deepwater Horizon spill, BP ranked fifth in the world in the value of non-state owned firms. With a market value of $233.2 billion, BP had been rising rapidly in both value and rank. It ranked second among oil firms, and followed only ExxonMobil, General Electric, Microsoft, and Citigroup, in that order.[139]

By May of 2008, its market capitalization remained over $206.7 billion. By the third week in June, 2010, BP met its nadir at $85.8 billion. In the span of just over a year, the company had lost almost $121 billion, or almost 60% of its value. By September of 2010, the stock value had rebounded somewhat, to a market capitalization of $120 billion.

With $20 billion set aside for economic damages, and with a potential worst case spill fine of $17.8 billion levied based on the Clean Water Act, the company has dropped in value by between two and three times the expected losses due to the spill. Even the worst case and harshest estimates of potential damages BP could face come nowhere near the drop in market value.

It is not difficult to trace the declines, and the partial rebound, in market value to specific events since the beginning of 2010.

Concentrating on the past year, with stock prices hovering between $55 and $60 per American Depository Receipts (ADR, where six ADRs represent one BP share), we see that stock prices dropped precipitously after the platform explosion, and continued to drop until CEO Tony Hayward was assigned away from management of the spill response and U.S. public relations.

Stocks also began to rebound when the six-month moratorium on offshore exploration, imposed by the Obama administration, was lifted by Federal Judge Feldman in Louisiana in the last week of June in 2010.

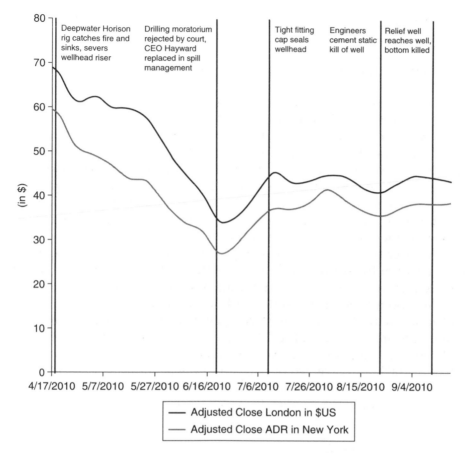

Figure 23.1 shows a graph with y-axis labeled "(in $)" ranging from 0 to 80, and x-axis dates from 4/17/2010 to 9/4/2010. Vertical lines mark events with the following labels across the top:

Deepwater Horison rig catches fire and sinks, severs wellhead riser

Drilling moratorium rejected by court, CEO Hayward replaced in spill management

Tight fitting cap seals wellhead

Engineers cement static kill of well

Relief well reaches well, bottom killed

Legend:
— Adjusted Close London in $US
— Adjusted Close ADR in New York

Figure 23.1 BP stock prices in the U.S. (New York) and England (London) as a function of spill events

Stocks continued to rise as BP made progress toward capping the well, and remained in a relatively tight trading range of around $40 per share ever since.

Specifically, we can observe one other phenomenon from the charts. ADRs are certificates that are denominated in U.S. dollars and issued by a bank that holds the equivalent amount of stock abroad. By exchanging these certificates rather than the stocks themselves, U.S. investors are freed of the duties and taxes that would otherwise complicate foreign stock purchases.

Normally, the ADR price should roughly track the underlying stock value, adjusted for the exchange rate between the underlying currencies, and the stock to certificate equivalency. In the case of BP ADRs, the following formula should define the equivalency:

Value of 1 BP stock = (1/6)×(pounds/dollar)×Value of one BP ADR

For instance, with a closing price, as of September 24, 2010, of £404.9, the BP ADR should be valued at $43.13, given the exchange rate of £0.6319 per dollar. However, the value of the ADR in New York was $38.46 at the close of the same day.

Upon closer inspection, of the chart, we see that the ADR had been lagging the true BP value, sometimes by as much as $15, or almost a 25% discount. This discount had narrowed, to about five dollars, or just over 10% in the weeks since the end of the spill and the removal of images of leaking oil from the media.

However, the U.S. ADR does not typically sell at a discount. For instance, if we look at the period one year before the spill, we see a reversal of the pattern at times, sometimes substantially so (Figure 23.2).

Nine months before the spill, the ADR was selling for $60, a premium of more than 30% over the sterling equivalent. The pattern reversed upon the dramatic fall of the U.S. and world stock markets in September of 2008, converged again as world markets began to recover in the latter half of 2009, and diverged once again in the beginning of 2010.

The divergence is most dramatic since the spill. This divergence demonstrates that the U.S. market depressed BP stock much more acutely, in

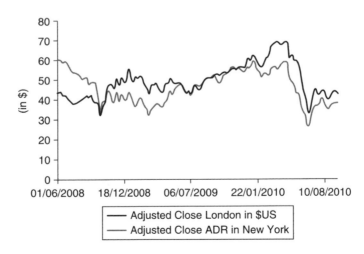

Figure 23.2   Weekly stock prices before the spill

unison with the harsher sentiment of BP among the U.S. media, public, and politicians.

An even longer look at BP stock will give some insight into the long-term convergence of BP stock prices. As part of the settlement with the Obama administration, BP agreed to place dividends in abeyance for 2010. The foregone dividend will amount to approximately $31.2 billion in foregone dividends. Even if the maximum fine was levied by the federal government based on the Clean Water Act, half of the $20 billion escrow fund for economic damages is prepaid, and the various emergency response and cleanup costs are paid in full, BP should have been able to pay these expenses from two quarters of foregone divi-dends in 2010. Any additional expenses to fund the final two years of the four-year economic damages escrow account schedule, and another incidental costs, could be funded by the tax advantage savings as the expenses are written off.

Consequently, a model of rational stock pricing for the BP equity should suggest BP would return to fundamental stock market valua-tions prevalent pre-spill and pre-recession, once the economy recovers and once BP is free to divert its sizeable income and cash flow to new investment and to dividends.

Figure 23.3 Weekly stock prices since 2003 in London and New York

This longer-term market valuation is shown by our final chart in Figure 23.3.

We see that the long-term price of BP stock in London has moved within a relatively tight range of between $40 and $50 since 2004–2005. It had risen above that range in the year preceding the oil spill, but, upon rebounding from the spill, remains in its relatively tight range. On the other hand, the U.S. market has demonstrated a greater optimism for BP shares since early 2004, except since the beginning of the recession, and especially since the spill.

With foregone dividends of approximately $4 billion per quarter, the total cost of the spill, economic claims, and fines would represent in the range of six quarters of payments. This suggests that by 2012 profitability could return, and BP stock could resume trading in the range of $50 to $60 per ADR. Indeed, by March of 2011, the ADR had already recovered to $47 per share.

Part VI

Where Do BP, Big Oil, and Energy-Starved Consumers Go from Here?

The Deepwater Horizon spill will have no redemption if we could not somehow learn from it. Obviously, BP has garnered a painful set of necessary practices. However, we will have lost an opportunity if the entire Big Oil industry does not learn a lesson equally profound. And, we as consumers cannot extricate ourselves from our shared responsibility. Like illicit drug exporting nations' refrain, there is no supply of drugs if there is no demand. Like drugs, oil is a demand driven industry, and we must better understand our role in this dynamic if we are to prevent many Deepwater Horizon scale spills in our energy future.

24
Reform of Regulatory Oversight

The Minerals Management Service, renamed the Bureau of Ocean Energy Management, Regulation, and Enforcement shortly after the spill, is a collection of 1,700 employees that regulate those industries which extract oil and minerals from the nation's outer continental shelf. It collects more than $13 billion per year from oil royalties. It has also been a lightning rod for controversy for years.

While other entities such as OSHA and the EPA had responsibility to ensure occupational and environmental safety and health, MMS was the lead agency responsible for inspection and oversight of the oil companies in the Gulf of Mexico. While its headquarters was in Washington, D.C., much of its activity was centered in the cozy circle of oil people in Houston, Texas, and Lake Charles, Louisiana.

The MMS does not have a long history. It was created through the passing of the Federal Oil and Gas Royalty Management Act by Congress in 1982. While it was originally charged to manage the nation's offshore hydrocarbon resources, its mandate was expanded in 2005 with the passage of the Energy Policy Act to include renewable energy resources on the Outer Continental Shelf. The MMS was renamed the Bureau of Ocean Energy Management, Regulation and Enforcement (BOEMRE) on June 21, 2010, just two months after the explosion on the Deepwater Horizon rig.

From its inception, but especially in the last two decades, the MMS was riddled with scandal. While scandals may point to the foibles of individual employees and may not represent an impairment of the organizational mission, the controversies at the MMS have placed squarely into doubt its ability to fulfill its mission. In particular, accusations of prostitution in 1990, and reports in 1998 that it comingled over drugs and sex with the energy company executives it regulated cast into doubt its ability to act as an impartial, stalwart, and responsible overseer of the offshore energy industry.[140]

The latter accusations, following a period in which reform attempts at the MMS had apparently failed, resulted from a thorough investigation from the Interior Department's Office of the Inspector General. It uncovered instances in which their Colorado office partied with and accepted gifts from the energy companies it regulated.

In response, immediately after the inauguration of President Obama, the newly appointed Department of the Interior Secretary Ken Salazar ordered a new code of conduct and an ethics reform program for the organization. Later that year, a bill was promulgated in Congress to reorganize the MMS because of its cozy relationship with the energy firms it was supposed to regulate may have caused the nation to forego billions of dollars of revenue.

Earlier in April, the month of the Deepwater Horizon explosion, Congress' General Accountability Office reported that the MMS in Alaska was invoked in lawsuits that claim it was shirking its responsibility to fully explore the environmental effects of hydrocarbon lease sales.

While the effectiveness of the regulator has been cast into question, it should not be deduced that problems in some quarters were shared by those inspecting rigs, overseeing oil companies, and enforcing safety and environmental regulations. However, within weeks of the Deepwater Horizon fire and spill, the Department of the Interior Secretary Salazar proposed splitting up the royalty and lease sale activities from those that regulate the exploration and development of offshore oil. A week later, he further proposed to separate exploration and production regulation, just as the director who oversaw oil and gas exploration at the MMS announced he would retire, effective within the month.

Meanwhile, the Interior Department's inspector general released another report, one month after Transocean's Deepwater Horizon platform sank, that accused the MMS's Lake Charles, Louisiana office of accepting significant gifts from the companies it oversaw. A couple of days later, Elizabeth Birnbaum, the MMS director resigned. In announcing her resignation, President Obama recognized that the MMS had not reformed "a culture in which oil companies were able to get what they wanted without sufficient oversight and regulation." In his first major policy speech following the spill, the President slammed the MMS and imposed a moratorium on new drilling until better permit approval protocols can be developed. He stated:

> In recent months, I've spoken about the dangers of too much – I've heard people speaking about the dangers of too much government

regulation. And I think we can all acknowledge there have been times in history when the government has overreached. But in this instance, the oil industry's cozy and sometimes corrupt relationship with government regulators meant little or no regulation at all.

When Secretary Salazar took office, he found a Minerals and Management Service that had been plagued by corruption for years – this was the agency charged with not only providing permits, but also enforcing laws governing oil drilling. And the corruption was underscored by a recent Inspector General's report that covered activity which occurred prior to 2007 – a report that can only be described as appalling. And Secretary Salazar immediately took steps to clean up that corruption. But this oil spill has made clear that more reforms are needed.

For years, there has been a scandalously close relationship between oil companies and the agency that regulates them. That's why we've decided to separate the people who permit the drilling from those who regulate and ensure the safety of the drilling.

I also announced that no new permits for drilling new wells will go forward until a 30-day safety and environmental review was conducted. That review is now complete. Its initial recommendations include aggressive new operating standards and requirements for offshore energy companies, which we will put in place.[141]

Obama went on to add:

What's also been made clear from this disaster is that for years the oil and gas industry has leveraged such power that they have effectively been allowed to regulate themselves. One example: Under current law, the Interior Department has only 30 days to review an exploration plan submitted by an oil company. That leaves no time for the appropriate environmental review. They result is, they are continually waived. And this is just one example of a law that was tailored by the industry to serve their needs instead of the public's. So Congress needs to address these issues as soon as possible, and my administration will work with them to do so.

The frustrations arose from the collapse of an oil rig that had been inspected only spottily in the previous years and from the release of a report from the Department of the Interior's Office of the Inspector General based on its investigations days before the Deepwater Horizon fire and spill. In the report's release on May 24, 2010, the Inspector

General presents a number of concerns that would directly question MMS's effectiveness in oversight of the Transocean rig and the BP drilling plan. The report noted:[142]

- Inspectors move with ease between government and industry jobs.
- Regulators fraternize and accept gifts from members of the energy industry that they have known for years, sometimes since childhood.
- The Outer Continental Shelf Act requires MMS to routinely inspect the 4,000 platforms of about 130 oil and gas companies operating in the Gulf. In five years, these platforms were fined a total of $572,500, or less than $1,000 per company per year.
- The MMS also issues annual "safe awards" to its oil companies.

It is not without irony, but nor does it suggest chicanery, that both Transocean and BP were recipients of the safe award.

On the other hand, a *Rolling Stone* magazine investigative article published on June 24, 2010 uncovered a series of problems with the MMS inspection and oversight protocol in the Gulf of Mexico. It noted:[143]

- The MMS was ill-equipped to test the assumptions made in oil company permit applications. While its jobs was to oversee and approve processes and permits, including modifications of permits, for instance, with regard to environmental analyses, its employees are technicians, not scientists, MMS agents. Notably, accepted BP's assertion that in the unlikely event of a spill at the Macondo Prospect, no adverse impacts wildlife was anticipated.
- Spill plans were not required to produce a site-specific plan in response to a blowout. Instead, the major oil companies invoke what turns out to be a generic "Oil Spill Response Plan" that, it turns out, is woefully inadequate and inappropriate for the region.

Another investigation by the Associated Press reported that:[144]

- In the five years before the explosion and fire, the Deepwater Horizon platform was inspected at least three times less each year than a previous policy mandated.
- Transocean had not received an infraction since 2003.
- In two instances in 2002, Transocean was cited with warnings or major safety violations regarding the effectiveness of their blowout preventer maintenance.

- A relaxation of safety requirements meant that rig operators themselves determine almost exclusively their design and implementation of safety processes.
- The Deepwater Horizon's safety record was so exemplary that it was exempted from inspectors' informal "watch list."

These reports paint a relatively consistent picture. The MMS may have been fully occupied regulating and inspecting problem explorers and producers, but it may have been less than diligent in overseeing those largest exploration entities. Instead, it may have deferred to the substantial engineering resources of these companies. Various people have called into question whether this lack of oversight was in deference to expertise, due to incompetence or a too-cozy relationship between regulators and Big Oil, or the product of outright corruption of the regulatory process.

However, while corruption or incompetence is episodic in any organizations, and widespread in few organizations, there is no claim that these problems influenced the permitting of BP and Transocean at the Macondo Prospect. We must await further investigation before we conclude that there were serious flaws in the permitting process that approved the drilling plans and modifications submitted by BP, or in maintenance procedures followed by Transocean. Meanwhile, MMS is gearing up for a stricter approval regime in light of the Deepwater Horizon spill and the resulting drilling moratorium.

25
What Do We Do with the World's Insatiable Need for Energy

Before we conclude, we must return to the root of the problem.

BP stands accused of failing to provide a sufficient number of barriers to the migration of hydrocarbons from the well bottom to the platform. The argument goes like this. If BP had better tested the Halliburton's cementing job, the accident would not have happened. If there were more barriers to migrating oil and gas, the failure of the cement would not have caused the accident. And, if BP or Transocean had properly interpreted problematic pressure readings in the hours before the explosion, the spill would not have occurred. If the rig crew had detected problems in the hour and minutes before the blowout, or vented hydrocarbon-laden mud overboard rather than into the gas-water separator, the accident likely would not have occurred. Finally, if the blowout preventer had worked as designed, the accident would not have happened.

This responsibility by inference should add one more qualification. If we had continued on a path of alternative energy production that was initiated more than 30 years ago under Jimmy Carter's unpopular presidency, we would not be forced to explore in such technically challenging and sensitive environments in the first place.

Unfortunately, while President Carter had narrowed America's energy dependency to just over four million barrels a day by the end of his decade, the United States disbanded most of his alternative energy programs. Most famously and symbolically, his successor, President Ronald Reagan, dismantled the solar cells President Carter had installed in the White House. More recently, the United States imports had risen to almost 14 million barrels per day. The United States is the world's largest oil importer, and there is no indication that this pattern will be reversed.

As a matter of fact, the rate of imported oil abated only modestly in the days of high oil prices in 2007 and 2008. U.S. oil demand is very robust and growing, as have been the profits of the big oil companies.

Further exacerbating this tendency is the greater competition for oil from emerging nations, as an earlier chapter documented. Demand for oil will increase dramatically as 2.5 billion people in China and India increasingly begin to demand the same level of consumption that their counterparts in the First Economic World enjoy. The gap between demand and supply will widen, and the price of oil will increase.

These increases will far exceed the ability of new production to keep up, and for new exploratory wells to identify new stocks of economic oil. Meanwhile, the price incentives to develop oil that, until now, has been impossible to reach will pay for the development of new drilling technologies, more complex systems, and greater environmental and technical challenges. This is an inevitability.

Equally inevitable is that greater technology offers greater opportunities for failure. And, more challenging environments mean that, should a failure occur, damages will be more severe and containment more difficult and time consuming.

We may have seen the end of easy oil.

We must instead acknowledge these challenges and create even more redundancies than we have relied upon in the past. These challenges we must master in the Gulf of Mexico are even more compounded when countries began to drill in even deeper waters off Brazil, in China, and in the Arctic Ocean.

We will also witness the increasingly desperate pursuit of offshore energy by countries without the same legal structure, liability rules, or regard for the environment. This too is inevitable, as we have already seen, given some of the environmental tragedies we are witnessing in Nigeria. We recognize that environmental concern is a luxury good. Our concern is increased as our income rises because we then have more to lose and we are willing to pay more for our own personal safety and to safeguard the environment. While the nations of the First Economic World have experienced a concomitant increase in income and regard for the environment over 60 years, other major nations are only beginning that cycle. The First Economic World can exercise little moral suasion to prevent aspiring nations from following the same oil-intensive path to economic development. Rather, we must, to some degree, wait for them to realize our same inconvenient truths. In the meantime, it seems inevitable that accidents will happen.

We can, and should, put into place stricter regulatory protocols, demand even higher technologies, and adopt international standards to prevent a Deepwater Horizon blowout from ever occurring again. However, there is only one antidote that could prevent another accident from happening. We must research and develop new and sustainable energy sources, and be willing to license and make these new technologies affordable to those in the emerging markets. And, when we drive our cars or heat our homes, we must acknowledge that the price of convenience or warmth is the spill in the Gulf of Mexico.

Assuming we remain wed to the cars and trucks that consume so much of our oil, what can we do to reduce our global demand for new and riskier petroleum exploration and production technologies?

First, the fabled hydrogen economy is no substitute at all. Without a source of extreme heat or electricity, we cannot produce hydrogen from water. And, as a lighter than air gas, most all the naturally occurring hydrogen has long since dispersed.

A hydrogen economy or a transportation sector fueled by batteries and electric motors still faces significant technological challenges. While these challenges are being addressed with impressive rapidity, we still require large-scale electric generation to make them work. There are only two options for such large-scale electric generation.

Coal power is a major generator of greenhouse gases, and its particulates harm the environment, in human and animal mortality, and in its economic consequences, at a rate much, much greater than created by the oil industry, even for oil in the risky deepwater locations that has been the focus of this book. However, the environmental and human costs of coal are less obvious to us than are oiled seabirds or images of a pipe spewing crude into the Gulf.

Alternately, we can develop fourth-generation designs for nuclear power plants that can contribute to a solution rather than be considered part of the problem.[145] However, nuclear power is almost as stigmatized as BP, and perhaps for good reason. The meltdown at Chernobyl and the inner containment failure at Three Mile Island invoked fears of nuclear bombs exploding in neighborhoods and trepidation about the unknown world of nuclear physics.

Meanwhile, a regulatory mechanism that essentially rewarded companies to invest in technologies that are highly capital-intensive, and then passed the costs and allowable profits on to consumers, encouraged high-cost technologies and cost overruns. Consumer revolt from both our environmental and our economic fears put a halt to applications for new nuclear power plants for decades. Meanwhile, newer plants of superior design have been built in Europe, India, and China.

Before we can begin to solve the nuclear-waste problem, and also tackle our growing energy needs, we must figure out how to navigate the public-opinion landmines. Other nations have figured out how to do this. France, that bastion of environmental sensitivity, produces almost all of its power from nuclear energy and is helping develop the new European third-generation designs. Meanwhile, nuclear power generates 16 of worldwide electricity needs and rids the environment of 2.5 billion tons of carbon dioxide greenhouse gases annually.

Our failure to have an educated public debate has actually imposed more risk than necessary on the American public. I hope we can begin to have an educated discussion over an admittedly complicated subject.

This is one example in which the United States could reestablish its lead in green energy development. There will not be a single silver bullet that solves its energy dilemma. But, with clear thinking and stronger leadership, it can take these U.S. innovations that are being employed in Canada and elsewhere and use them to benefit the American public. To do so, though, it will need to overcome the shrill debate over all forms of new and cleaner energy. Ultimately, the world needs an educated and productive public discussion of all possible ways to increase energy self-reliance.

It is clear that our choices will be constrained and difficult. Until we learn to accept some of the responsibility for the technologies we employ and the dangers they create, we must become more informed and engage in some important public debates. The shrill whine of politics and the demonizing of actors, even bad actors, merely deflects our personal responsibility and delays an eventual solution.

26
Conclusion

It is difficult to argue that there is such a thing as a prevalent cross-corporate culture that would lead a company such as BP to risk the lives of its workers. All will agree that mistakes were made. However, while companies may have pockets of operations that must become more diligent in its safety procedures, no corporation as large as BP can be so easily caricatured.

In this book, I took a close look at one BP operation – the Deepwater Horizon spill in the Gulf of Mexico. This exploration culture is very different than the culture of exploration in Alaska, the North Sea, and elsewhere, is certainly distinct from the culture in production operations, and shares little with the culture found at refineries, or gas stations, for that matter.

The exploration culture is one of high technology and high level engineering. It is a rarified culture in which only the best of the best can successfully and routinely engineer wells that would have been considered next to impossible just a couple of decades ago. It is actually a culture with an excellent safety record, primarily because of the level of mechanization and high technology necessitated by the extreme technologies it must master every day.

Certainly, technology and management failed this time and in a most environmentally sensitive and visible setting. Surely we will discover that various other companies or individuals played a role in the failure that is most closely associated with BP. However, this cannot be solely about oil companies, despite our best efforts to simplify a most complex problem.

The exploration culture has perhaps become complacent, given the success of deep-sea exploration over the past decade. Now, it is also a culture that cannot afford to not learn from its mistakes, and will change its practices and protocols, industry-wide, based on what it has learned from this calamity. Deep-sea oil exploration is a process we all now depend upon, and will even more so even in the more immediate future.

I have no doubt that, in the final analysis, there will be plenty of blame to go around, from BP to Transocean, Halliburton to Cameron International, the Minerals Management Service to all of us that demand Big Oil to explore in more difficult and less repairable environments.

When we instead shirk this latter responsibility and consider Big Oil as a monolithic and immoral entity that earns huge profits and usurps our income as it exploits earth's resources, our consciences may be temporarily appeased. However, our own ignorance only makes the problems, and their solutions, more intractable.

These companies have been growing at an average double-digit rate per year and are now dominating global commerce because we increasingly demand their products. As they grow and take advantage of greater economies of scale, their problems are magnified. No longer do we see a series of accidents in ten companies, but we see ten accidents in one large and monolithic company. And so, it appears to us that an industry that is actually getting safer appears to have become more reckless.

Of course, with their size inevitably come some resentment, and the possibility of politics. Certainly, when politics looms, the mischief of invoking fears of foreign influence cannot help but stoke a nation's other fears as well.

Ultimately, though, we have the power to better understand what we are demanding, the risks that will result, and the ways in which modern organizations function as they grow. We may, as a consequence of a healthier public debate, demand changes from ourselves and from Big Oil. However, if we instead go on the offensive and force Big Oil to go on the defensive, useful discussion stops as rapidly as we shirk our collective responsibility. We have the power to get the debate back on track if we are able to discuss complex realities without simple appeal to political histrionics. That is our challenge, just as it is expected that BP will do the best possible job to live up to their challenges.

Notes

1. http://www.gomr.boemre.gov/homepg/lsesale/206/pstat206.pdf, accessed October 22, 2010.
2. Broad, William, J., "Tracing Oil Reserves to Their Tiny Origins," *The New York Times*, August 2, 2010, http://www.nytimes.com/2010/08/03/science/03oil.html?_r=1&pagewanted=print, accessed August 19, 2010.
3. Barnola, J.-M., D. Raynaud, C. Lorius, and N.I. Barkov (2003). "Historical CO2 record from the Vostok ice core." In *Trends: A Compendium of Data on Global Change*. Carbon Dioxide Information Analysis Center, Oak Ridge National Laboratory, U.S. Department of Energy, Oak Ridge, Tenn., U.S.A.
4. http://www.esrl.noaa.gov/gmd/ccgg/trends/, accessed October 22, 2010.
5. "The Role of Deepwater Production in Global Oil Supply," Cambridge Energy Research Associates, June 30, 2010, http://press.ihs.com/article_display.cfm?article_id=4267, accessed August 19, 2010.
6. http://www.eia.doe.gov/emeu/international/reserves.html, accessed October 22, 2010.
7. One mole is equal to 6.022×10^{23}. A mole is a conversion factor that represents the number of molecules that would have a mass in grams equal to the sum of the mass of elements, in atomic weight units, that make up the molecule. For instance, a mole of methane molecules would weigh 16 grams, the sum of a mole of carbon with an atomic weight of 12, and four moles of hydrogen atoms with a weight of one atomic unit.
8. http://en.wikipedia.org/wiki/Petroleum, accessed September 2, 2010.
9. The 33 countries of the Organization for Cooperation and economic Development are Australia, Austria, Belgium, Canada, Chile, Czech Republic, Denmark, Finland, France, Germany, Greece, Hungary, Iceland, Ireland, Israel, Italy, Japan, Korea, Luxembourg, Mexico, Netherlands, New Zealand, Norway, Poland, Portugal, Slovak Republic, Slovenia, Spain, Sweden, Switzerland, Turkey, United Kingdom, and the United States.
10. http://www.eia.doe.gov/oiaf/ieo/graphic_data_world.html, accessed October 3, 2010.
11. http://www.upi.com/Science_News/Resource-Wars/2009/09/03/Worlds-deepest-well-taps-giant-oil-find-in-the-US-Gulf-of-Mexico/UPI-22801252018451, accessed September 1, 2010.
12. Harvey, Steve, "California's Legendary Oil Spill," *Los Angeles Times*, June 13, 2010.
13. Rintoul, William, *Spudding In: Recollections of Pioneer Days in the California Oil Fields*. San Francisco: California Historical Society, 1976, pp. 106–113. ISBN 0910312370.
14. http://ludb.clui.org/ex/i/CA3068/, accessed August 16, 2010.

15. "The Lakeview Gusher," http://web.archive.org/web/20061019100520/ http://www.sjgs.com/lakeview.html, retrieved October 2, 2010.
16. Baumann, Paul, "Environmental Warfare: 1991 Persian Gulf War," manuscript from the Department of Geography, State University of New York College at Oneonta, 2001, http://employees.oneonta.edu/baumanpr/ geosat2/Environmental_Warfare/ENVIRONMENTAL_WARFARE.htm, accessed August 16, 2010. U.S. Senate Committee on Environment and Public Works. 1992. "The Environmental Aftermath of the Gulf War Congressional Report," p. 74.
17. "Gulf Found to Recover From War's Oil Spill, *The New York Times*, March 18, 1993, http://www.nytimes.com/1993/03/18/world/gulf-found-to-recover-from-war-s-oil-spill.html, accessed August 16, 2010.
18. Linda Garmon (25 October 1980). "Autopsy of an Oil Spill". *Science News* 118 (17), pp. 267–270.
19. http://en.wikipedia.org/wiki/Oil_spills#Largest_oil_spills, accessed August 17, 2010.
20. ERCO/Energy Resources Co. Inc., 19 March 1982, Ixtoc Oil Spill Assessment, Final Report, Executive Summary Prepared for the US Bureau of Land Management, Contract No. AA851-CTO-71. US Department of the Interior, Minerals Management Service Mission, page 27, http://invertebrates.si.edu/ mms/reports/IXTOC_exec.pdf, accessed August 17, 2010.
21. http://www.cedre.fr/en/spill/atlantic/atlantic.php, accessed August 17, 2010.
22. http://www.nationmaster.com/graph/ene_oil_con-energy-oil-consumption, accessed August 17, 2010
23. http://www.itopf.com/information-services/data-and-statistics/case-histories/alist.html, accessed August 17, 2010.
24. http://www.incidentnews.gov/incident/6262, accessed August 17, 2010.
25. Moldan, A.G.S., L.F. Jackson, S. McGibbon, and J. Van Der Westhuizen. "Some Aspects of the Castillo De Bellver Oilspill," *Marine Pollution Bulletin* 16 (3), March 1985, pp. 97–102.
26. http://www.nysun.com/new-york/greenpoint-maspeth-residents-lobby-to-get-55-year/23231/, accessed August 17, 2010.
27. 2005 Essay from the Santa Barbara Wildlife Care Network, http://www2. bren.ucsb.edu/~dhardy/1969_Santa_Barbara_Oil_Spill/About.html, accessed October 12, 2010.
28. bp Plans to Pull Another 2b Barrels of Oil from Alaska's Prudhoe Bay. Rigzone. February 22, 2008, http://www.rigzone.com/news/article.asp?a_ id=57255, accessed August 19, 2010.
29. The two temperature scales intersect at the temperature −40 degrees.
30. Leveson, Nancy, "Software System Safety," Massachusetts Institute of Technology class notes, http://ocw.mit.edu/courses/aeronautics-and-astronautics/16–358j-system-safety-spring-2005/lecture-notes/class_notes. pdf, retrieved August 19, 2010.
31. Hunter, Don, "Alcohol Stains Record of Skilled Sea Captain," *Anchorage Daily News*, March 24, 1989, http://www.adn.com/evos/stories/EV53.html, accessed August 19, 2010.

32. Bluemink, Elizabeth, "Size of Exxon spill remains disputed," *Anchorage Daily News*, June 10, 2010, http://www.adn.com/2010/06/05/1309722/size-of-exxon-spill-remains-disputed.html, accessed August 19, 2010.

33. Holleman, Marybeth, "The Lingering Lesson of the Exxon Valdez Spill," *The Seattle Times*, March 22, 2004, http://www.commondreams.org/views04/0322-04.htm, accessed August 19, 2010.

34. Graham, Sarah, "Environmental Effects of Exxon Valdez Spill Still Being Felt," Scientific American, December 19, 2003, http://www.scientificamerican.com/article.cfm?id=environmental-effects-of, accessed August 19, 2010.

35. "An Assessment of the Impact of the Exxon Valdez Oil Spill on the Alaska Tourism Industry." Prepared by Preston, Thorgrimson, Shidler, Gates, Ellis, and the McDowell Group for the Exxon Valdez Oil Spill Trustee Council, http://www.evostc.state.ak.us/Universal/Documents/Publications/Economic/Econ_Tourism.pdf, accessed August 19, 2010.

36. Carrigan, Alison C., "The Bill of Attainder Clause: A New Weapon to Challenge the Oil Pollution Act of 1990," *Boston College Environmental Law Review*, 2000–01, http://www.bc.edu/bc_org/avp/law/lwsch/journals/bcealr/28_1/04_FMS.htm, accessed August 19, 2010.

37. "The Recently Negotiated Settlement of Civil and Criminal Liabilities Resulting from the Exxon Valdez Oil Spill," Hearing before the Committee on Merchant Marine and Fisheries, United States House of Representatives, 102nd Congress, Congressional Research Service, Library of Congress, Serial No. 102–8, March 20, 1991

38. http://iml.jou.ufl.edu/projects/spring01/hogue/exxon.html#ineffective, accessed September 3, 2010.

39. Elliott, Stuart, "Exxon's Image Soiled; Public Angry at Slow Action on Oil Spill," *USA Today*, April 21, 1989.

40. "Exxon Is Weaker Than You Think," *Business Week*, February 12, 2009, http://www.zimbio.com/BusinessWeek/articles/693/EXXON+IS+WEAKER+THAN+YOU+THINK, accessed September 3, 2010.

41. Schmit, Julie, "bp Could Learn from Exxon's Safety Response to Valdez Oil Spill," *USA Today*, August 4, 2010, http://www.usatoday.com/money/industries/energy/2010-08-03-oilspillrig_N.htm, accessed October 22, 2010.

42. http://www.boemre.gov/ooc/press/2010/press0903.htm, accessed October 25, 2010.

43. http://en.wikipedia.org/wiki/Piper_Alpha, accessed October 24, 2010.

44. http://www.redadair.com/hisstory.html, accessed October 24, 2010.

45. http://www.blogofdeath.com/archives/001119.html, accessed October 24, 2010.

46. Commission report, p. 70.

47. Commission report, p. 71.

48. Red Adair was a pioneer in the containment of large, flaming gushers. He often used explosives to collapse the wellhead and to rob the flame of the oxygen it needed to support combustion.

49. Lochhead, Carolyn, "San Bruno Fire Leads to Vow of Safety Review," SFGate. com, September 29, 2010, http://www.google.com/search?q=san+bruno+ex plosion+killed&rls=com.microsoft:en-us&ie=UTF-8&oe=UTF-8&startIndex= &startPage=1#sclient=psy&hl=en&rls=com.microsoft:en-us&tbs=nws:1%2C sbd%3A1&q=san+bruno+explosion+killed+smelled+gas&aq=f&aqi=&aql=& oq=&gs_rfai=&pbx=1&fp=d33939b9cb55eb65, accessed October 4, 2010.
50. Barrish, Robert A. (July 23, 2001). "Taking the Hazard Out of Hazardous Chemicals." Division of Air and Waste Management – Air Quality Management, http://web.archive.org/web/20050907185634/http://www. dnrec.state.de.us/air/aqm_page/brabson.htm, retrieved September 3, 2010.
51. The database can be found at: http://www.osha.gov/pls/imis/industry.html, last accessed March 23, 2011.
52. http://www.osha.gov/pls/imis/sic_manual.display?id=389&tab=description, accessed September 3, 2010.
53. http://www.boemre.gov/incidents/fatalities.htm, accessed September 4, 2010.
54. http://www.usatoday.com/money/industries/energy/2010-08-03-oilspillrig_N.htm, accessed September 4, 2010.
55. http://www.bp.com/extendedsectiongenericarticle.do?categoryId=9033194 &contentId=7060828#tab3, accessed September 4, 2010.
56. http://energycommerce.house.gov/documents/20100615/ transcript.06.15.2010.ee.pdf, accessed September 4, 2010.
57. http://www.businessweek.com/news/2010-06-07/drilling-rules-change-may-harm-industry-exxon-says-update1-.html, accessed September 4, 2010.
58. http://www.subseaiq.com/data/Project.aspx?project_id=562&AspxAuto, accessed September 4, 2010
59. "Debottlenecking Removes Auger Production Constraints," *Oil and Gas Journal* 94 (6), November 11, 1996.
60. *Deep Water – The Gulf Oil Disaster and the Future of Offshore Drilling,* Report to the President of the National Commission on the BP Deepwater Horizon Oil Spill and Offshore Drilling, January 12, 2011, p. 38.
61. Commission report, p. 50.
62. http://www.whitehouse.gov/the-press-office/remarks-president-a-discussion-jobs-and-economy-charlotte-north-carolina, accessed October 10, 2010.
63. http://www.subseaiq.com/data/Project.aspx?project_id=562&AspxAuto, accessed September 4, 2010
64. Washburn, Mark, "a Huff and Boom Ended Deepwater Horizon's Luck," *The McClatchy Company,* May 14, 2010, http://www.mcclatchydc. com/2010/05/14/94184/a-huff-and-boom-ended-deepwater.html, accessed September 4, 2010.
65. http://www.deepwater.com/_filelib/FileCabinet/fleetupdate/2010/RIGFLT-APR-2010.xls?FileName=RIGFLT-APR-2010.xls, accessed September 4, 2010.
66. Callelman, Ben, "Rig Owner Had Rising Tally of Accidents," *Wall Street Journal,* May 10, 2010, http://online.wsj.com/article/SB100014240527487 04307804575234471807539054.html, accessed September 17, 2010.

67. Gold, Russell, and Ben Casselman, "Drilling Process Attracts Scrutiny in Rig Explosion," *Wall Street Journal*, April 30, 2010, http://online.wsj.com/article/SB10001424052748703572504575214593564769072.html, accessed September 17, 2010.
68. http://www.gomr.boemre.gov/homepg/offshore/safety/acc_repo/2006/060212.pdf, accessed September 17, 2010.
69. Hammer, David, "Deepwater Horizon Supervisor Confirms Leaks in Blowout Preventer, Lack of Certification," *New Orleans Times-Picayune*, August 25, 2010, http://www.nola.com/news/gulf-oil-spill/index.ssf/2010/08/deepwater_horizon_supervisor_c.html, accessed October 4, 2010.
70. Judge, Bob, and Gary Leach, "Subsea Test Valve in Modified bop Cavity May Help to Minimize Cost of Required bop Testing," *Drilling Magazine*, November/December, 2006, http://www.drillingcontractor.org/dcpi/dc-novdec06/DC_Nov07_judge.pdf, accessed October 4, 2010.
71. http://www.deepwaterinvestigation.com/external/content/document/3043/1047291/1/DNV%20Report%20EP030842%20for%20BOEMRE%20Volume%20I.pdf, accessed March 23, 2011.
72. http://energycommerce.house.gov/documents/20100614/BP-April14.Email.calling.Macondo.a.nightmare.well.pdf, accessed September 17, 2010.
73. *Deep Water – The Gulf Oil Disaster and the Future of Offshore Drilling*, Report to the President of the National Commission on the BP Deepwater Horizon Oil Spill and Offshore Drilling, January 12, 2011, pp. 4–19.
74. Commission report, p. 5.
75. Commission report, p. 6.
76. Commission report, p. 6.
77. Commission report, p. 8.
78. Hammer, David, "Oil Spill Hearings: Transocean Lawyer Disputes 2008 Accident Report," *New Orleans Times-Picayune*, May 28, 2010, http://www.nola.com/news/gulf-oil-spill/index.ssf/2010/05/oil_spill_hearings_transocean.html, accessed October 4, 2010.
79. http://energycommerce.house.gov/documents/20100614/Hayward.BP.2010.6.14.pdf, accessed September 17, 2010.
80. http://energycommerce.house.gov/documents/20100614/BP-March25.Email-long.string.saves.time.pdf, accessed September 17, 2010.
81. http://energycommerce.house.gov/documents/20100614/HAL-Production.Casing.Design.Report.4.15.2010.moderate.pdf, accessed September 17, 2010.
82. http://energycommerce.house.gov/documents/20100614/BP-April16.Email.exchange.about.centralizers.pdf, accessed September 17, 2010.
83. Commission report, p. 97.
84. Casselman, Ben, "Halliburton Engineer Says He Warned bp Seal Might Fail," *Wall Street Journal*, August 25, 2010, http://online.wsj.com/article/SB10001424052748703447004575450110807492410.html, accessed September 17, 2010.

85. Brown, Robbie, "Adviser Says He Raised Concerns to BP on Well," *The New York Times,* August 24, 2010. http://www.nytimes.com/2010/08/25/us/25hearings.html, accessed September 17, 2010.
86. http://www.oilspillcommission.gov/letter/letter-mr-bartlit-oil-spill-commission, accessed October 31, 2010.
87. 30 CFR 250.428
88. BP GOM Exploration Wells, MC252 #1ST00PBP01 – Macondo Prospect 7"x9-7/8" Interval, Rev. H.2, page 6, April 15, 2010.
89. "Deepwater Horizon Accident Investigation Report," September 8, 2010, http://www.bp.com/sectiongenericarticle.do?categoryId=9034902&contentId=7064891, accessed September 8, 2010.
90. "National Academy of Engineering Investigative Hearing," September 26, 2010, http://www.nae.edu/Activities/20676/deepwater-horizon-analysis.aspx, accessed October 2, 2010.
91. http://uanews.ua.edu/2010/05/ua-in-the-news-may-26–2010/, accessed September 18, 2010
92. Commission report, p. 124.
93. Pilkington, Ed., "Deepwater Horizon Alarms Were Turned Off 'to Help Workers Sleep'," *Guardian,* July 23, 2010, http://www.guardian.co.uk/environment/2010/jul/23/deepwater-horizon-oil-rig-alarms, accessed September 18, 2010.
94. http://www.empr.gov.bc.ca/OG/offshoreoilandgas/Pages/OffshoreOilandGasAroundtheWorld.aspx, accessed September 18, 2010.
95. For instance, see the BowTieXP software, as described by http://www.bow-tiexp.com.au/bowtiexp.asp#aboutBowTies, accessed September 18, 2010.
96. http://www.montarainquiry.gov.au/downloads/DirectionOrders/NPO%2019%20March%202010.pdf, accessed September 18, 2010.
97. Kirby, James, "Why the Oil Minister Doesn't Feel Too Well Right Now,´*Business Day,* June 20, 2010, http://www.theage.com.au/business/why-the-oil-minister-doesnt-feel-too-well-right-now-20100619-ynzc.html, accessed September 18, 2010.
98. http://www.iso.org/iso/iso_catalogue/catalogue_tc/catalogue_detail.htm?csnumber=31534, accessed September 19, 2010.
99. Covello, Vincent T.; and Frederick H. Allen (April 1988). *Seven Cardinal Rules of Risk Communication.* Washington, DC: U.S. Environmental Protection Agency. OPA-87-020.
100. http://www.epa.gov/pubinvol/pdf/risk.pdf, accessed September 18, 2010.
101. http://www.chevron.com/chevron/pressreleases/article/07212010_newoilspillcontainmentsystemtoprotectgulfofmexicoplannedbymajoroil-companies.news, accessed September 19, 2010.
102. http://www.chacha.com/question/how-many-drops-of-water-will-fit-in-a-five-gallon-bucket, accessed September 19, 2010.
103. "Modest Decline in Oil Leak Interest, Sharp Decline in Coverage," *Pew Research Center for the People and the Press,* July 14, 2010, http://people-press.org/report/634/, accessed September 19, 2010.

104. http://www.journalism.org/analysis_report/public_had_huge_appetite_ story, accessed September 19, 2010.
105. Glassner, Barry (1999). *The Culture of Fear*. New York: Basic Books.
106. http://www.epa.gov/pubinvol/pdf/risk.pdf, accessed September 18, 2010.
107. http://en.wikipedia.org/wiki/British_Airways_Flight_9, accessed September 19, 2010.
108. http://www.cleangulf.org/program.php, accessed October 25, 2010.
109. http://m.whitehouse.gov/blog/2010/05/14/relentless-efforts-stop-leak-and-contain-damage, accessed October 23, 2010.
110. http://www.bloomberg.com/news/2010-06-19/anadarko-says-bp-was-reckless-looks-to-oil-company-to-pay-spill-claims.html, accessed September 19, 2010.
111. Swint, Brian, "BP Says Spill Response 'Aggressive,' Well to Cost $100 Million," *Business Week*, April 27, 2010, http://www.businessweek.com/news/2010-04-27/bp-says-spill-response-aggressive-well-to-cost-100-million.html, accessed September 19, 2010.
112. http://uk.ibtimes.com/articles/20100504/bp-ramps-up-us-oil-spillage-response_all.htm, accessed September 19, 2010.
113. Poll Shows Negative Ratings for BP, Federal Government, *Washington Post*, June 7, 2010, http://voices.washingtonpost.com/behind-the-numbers/2010/06/poll_shows_negative_ratings_fo.html, accessed September 19, 2010.
114. Chen, Edwin, and Julianna Goldman, "Obama Outraged Over BP Spill, Federal Oil Regulators," *Businessweek*, May 27, 2010, http://www.businessweek.com/news/2010-05-27/obama-outraged-over-bp-spill-federal-oil-regulators-update1-.html, accessed September 20, 2010.
115. http://www.dailymail.co.uk/news/worldnews/article-1283672/GULF-OIL-SPILL-James-Camerons-view-BP-fix-fails.html, accessed September 24, 2010.
116. http://www.noaanews.noaa.gov/stories2010/PDFs/OilBudget_description_%2083final.pdf, accessed September 24, 2010.
117. Camilli, Richard, Christopher M. Reddy, Dana R. Yoerger, Benjamin A. S. Van Mooy, Michael V. Jakuba, James C. Kinsey, Cameron P. McIntyre, Sean P. Sylva, and James V. Maloney, "Tracking Hydrocarbon Plume Transport and Biodegradation at Deepwater Horizon," *Science* Online, August 19, 2010, http://www.sciencemag.org/cgi/content/abstract/science.1195223, accessed September 24, 2010.
118. Hazen, Terry, Eric Dubinsky, Todd DeSantis, Gary Andersen, Yvette Piceno, Navjeet Singh, Janet Jansson, Alexander Probst, Sharon Borglin, Julian Fortney, William Stringfellow, Markus Bill, Mark Conrad, Lauren Tom, Krystle Chavarria, Thana Alusi, Regina Lamendella, Dominique Joyner, Chelsea Spier, Jacob Baelum, Manfred Auer, Marcin Zemla, Romy Chakraborty, Eric Sonnenthal, Patrik D'haeseleer, Hoi-Ying Holman, Shariff Osman, Zhenmei Lu, Joy Van Nostrand, Ye Deng, Jizhong Zhou and Olivia Mason, "Deep-sea oil plume enriches Indigenous oil-degrading bacteria," *Science*, August 26, 2010.

119. http://www.sciencedaily.com/releases/2010/08/100824132349.htm, accessed September 24, 2010.
120. http://www.lmrk.org/corexit_9500_uscueg.539287.pdf, accessed September 24, 2010.
121. Yarris, Lynn, "Study Shows Deepwater Oil Plume in Gulf Degraded by Microbes," Berkeley Lab News Center, August 24, 2010, http://newscenter.lbl.gov/news-releases/2010/08/24/deepwater-oil-plume-microbes/, accessed October 23, 2010.
122. http://www.guardian.co.uk/business/2010/may/13/bp-boss-admits-mistakes-gulf-oil-spill, accessed September 24, 2010.
123. http://www.epa.gov/gmpo/about/facts.html, accessed September 24, 2010.
124. http://www.fws.gov/home/dhoilspill/pdfs/collection_08132010.pdf, accessed September 24, 2010.
125. http://www.evostc.state.ak.us/facts/qanda.cfm, accessed September 24, 2010.
126. Presidential commission, p. 210.
127. http://www.oilspillcommission.gov/sites/default/files/documents/Working%20Paper.Amount%20and%20Fate.For%20Release.pdf, accessed October 11, 2010.
128. Broder, John M., "Investigator Finds No Evidence That BP Took Shortcuts to Save Money," *The New York Times*, November 8, 2010, http://www.nytimes.com/2010/11/09/us/09spill.html, accessed November 9, 2010.
129. "The Amount and Fate of the Oil," *The National Commission on the BP Deepwater Horizon Oil Spill and Offshore Drilling*, http://www.fws.gov/migratorybirds/RegulationsPolicies/mbta/taxolst.html, accessed September 25, 2010.
130. http://www.fws.gov/endangered, accessed September 25, 2010.
131. Oil Spill Cost and Reimbursement Fact Sheet, http://www.restorethegulf.gov/release/2010/10/13/oil-spill-cost-and-reimbursement-fact-sheet, accessed October 23, 2010.
132. De los Rios, Alejandro, "bp Declares It Will Waive $75 opa Cap,", The Louisiana Record, October 19, 2010, http://www.louisianarecord.com/news/230560-bp-declares-it-will-waive-75-opa-cap, accessed October 23, 2010.
133. http://research.stlouisfed.org/fred2/series/UMCSENT/downloaddata?cid=98, accessed October 24, 2010.
134. http://www.zsz.com/new%20security%20cases.htm#bpgulf, accessed September 25, 2010.
135. Leary, Alex, "Obama Administration Directs Criticism to bp Claims Czar Kenneth Fienberg," *St. Petersburg Times*, September 25, 2010, http://www.tampabay.com/news/politics/stateroundup/obama-administration-directs-criticism-to-bp-claims-czar-kenneth-feinberg/1123951, accessed September 25, 2010.
136. Claims and Government Payments – Gulf of Mexico Oil Spill, Public Report, September 23, 2010, published by BP, http://www.bp.com/liveassets/bp_internet/globalbp/globalbp_uk_english/incident_response/

STAGING/local_assets/downloads_pdfs/Public_Report_9_23_2010.pdf, accessed September 25, 2010.

137. Feeley, Jef, and Margaret Cronis Fisk, "BP Gulf-Spill Lawsuits Consolidated in New Orleans," Bloomberg, August 10, 2010, http://www.bloomberg. com/news/2010-08-10/bp-gulf-oil-spill-lawsuits-to-be-consolidated-in-new-orleans-federal-court.html, accessed September 25, 2010.

138. Crone, Timothy J. and Maya Tolstoy, "Magnitude of the 2010 Gulf of Mexico Oil Leak," *Science* Online, September 23, 2010, http://www. sciencemag.org/cgi/content/abstract/science.1195840, accessed September 25, 2010.

139. Ranked by the *Financial Times*, http://media.ft.com/cms/b970931e-c2f8-11da-a381-0000779e2340.pdf, accessed September 25, 2010.

140. Eilperin, Juliet, and Madonna Lebling, "mms's Troubled Past," *The Washington Post*, May 29, 2010, http://www.washingtonpost.com/wp-dyn/content/article/2010/05/28/AR2010052804599_pf.html, accessed September 26, 2010.

141. White House transcript of Presidential Speech and Press Conference, May 27, 2010, http://www.whitehouse.gov/the-press-office/remarks-president-gulf-oil-spill, accessed September 26, 2010.

142. http://www.doioig.gov/images/stories/reports/pdf//IslandOperatingCo. pdf, accessed September 26, 2010.

143. Dickinson, Tim, "The Spill, the Scandal and the President," *Rolling Stone*, June 24, 2010, http://www.rollingstone.com/politics/news/17390/111965?RS_show_page=0, accessed September 26, 2010.

144. Associated Press, "Review: Oil Rig Inspections Fell Short of Guidelines," *Kingsport Tennessee Times News*, May 16, 2010, http://www.timesnews.net/article.php?id=9023118, accessed September 26, 2010.

145. Read, Colin, "Nuclear Power Needs Another Look," *Plattsburgh Press Republican*, September 19, 2010, http://pressrepublican.com/0216_read/x1967876178/Nuclear-power-needs-another-look, accessed September 26, 2010.

Glossary

Alkanes – simple saturated hydrocarbons, also called paraffins, that constitute the largest part of crude oil.

Alkenes – a saturated hydrocarbon in crude oil that contains a double carbon bond.

Allen, Retired Admiral Thad – the administrator appointed by President Obama to head the Unified Command System to coordinate the containment and cleanup of the Deepwater Horizon Oil spill.

Anadarko – a major Gulf of Mexico oil company and 25% owner of the Macondo Prospect, along with BP and Mitsui Oil Exploration.

Annulus – the space between a drill/production pipe and the casing (annulus -A) or the casing and the rock and sand surrounding the wellbore.

Automatic Mode Function (AMF) – a system that automatically activates blow-out preventer rams to shut down a well if other manual systems fail.

Bill of Attainder – a law designed to constrain an individual or sole organization.

Birnbaum, Minerals Management Service Director Elizabeth – the beleaguered head of the MMS that suddenly resigned in the aftermath of the Deepwater Horizon spill.

Bladder effect – a theory postulated by the Deepwater Horizon drill team to explain unusual negative pressure readings.

Blowout – the sudden and uncontrolled release of gas and oil from a well.

Blowout preventer (BOP) – a device that activates rams to pinch down a well pipe in the case of a blowout emergency.

Bottom kill – the cement sealing of a well near its bottom to prevent migration of gas and oil through the outer annulus between the wellbore and casing, the inner annulus between the casing and pipe, or within the pipe.

BP – a British limited liability corporation, formerly named British Petroleum, that is the 65% owner of the Macondo Prospect.

British Petroleum – the former name for BP.

Brown, Gordon – the Prime Minister of England at the onset of the Deepwater Horizon oil spill

Bureau of Ocean Energy Management, Regulation, and Enforcement – formerly known as the Minerals Management Service, the renamed agency under the Department of the Interior that is responsible for offshore oil lease sales and oversight.

Cameron International – the manufacturer of the blowout preventer that failed to seal the Macondo Prospect well.

Cameron, David – the Prime Minister of England at the conclusion of the Deepwater Horizon oil spill.

Cap – a cement or metal seal at the top of a well.

Casing – the steel lining that isolates the well pipe from the surrounding rock and sand.

Cement – a complex mixture that can be inserted as a liquid into a well hole and will subsequently harden to seal well gaps. Cement is differentiated from concrete, which also includes stone.

Chevron – one of the major oil companies that explore and produce in the Gulf of Mexico.

Chu, Energy Secretary Steven – a physicist that held the position of Secretary of Energy in the Obama administration during the Deepwater Horizon spill and its containment and remediation.

Civil Law – statutes or principles that govern claims between individuals or groups. Civil sanctions can include compensatory and punitive damages, but cannot invoke a jail sentence.

Clean Water Act – a series of acts that allow for fines for the discharge of oil or other contaminants into water resources in the United States.

ConocoPhillips – one of the major oil companies that explore and produce in the Gulf of Mexico

Cracking – the separation of crude oil into its constituent hydrocarbon parts.

Criminal – statutes or principles that govern claims between individuals or groups and society. Criminal sanctions can include compensatory and punitive damages, fines, and jail sentences.

Damages – the financial claims for loss of income, wealth, or other assets or other quantifiable losses.

Deepwater Horizon – the Transocean-owned and operated semisubmersible oil drilling rig that was commissioned by BP to drill the Macondo Prospect.

Department of Justice (DoJ) – the entity of the U.S. Government responsible for representing the people in suits against other parties.

Deterrent – a mechanism designed to deter actions not in the public interest.

Drill – a large, typically diamond-headed bit used to penetrate rock and sand in pursuit of oil.

Dudley, Bob – the BP executive who replaced Tony Hayward as BP's Chief Executive Officer in the aftermath of the Deepwater Horizon spill

Economic damages – quantifiable financial damages suffered as a consequence of an act of another.

Economic Oil – oil reserves that can be profitably extracted at prevailing oil prices.

Endangered Species Act – a series of acts that provide for fines for the killing of animals on the endangered species list.

Environmental Protection Agency – the lead U.S. government agency charged with protecting the environment.

Exxon Valdez – the most public oil spill disaster in the United States prior to the Deepwater Horizon spill.

ExxonMobil – one of the major oil companies that explore and produce in the Gulf of Mexico

Feinberg, Kenneth – a lawyer and frequent special master employed by the Obama administration to administer and streamline economic claims resolution in the aftermath of the Deepwater Horizon oil spill.

Flapper valve – a one way valve located in the well shoe designed to prevent oil from migrating to the surface in an exploratory well.

Gulf of Mexico – an oil rich and environmentally sensitive ocean gulf located along the shores of Mexico, the U.S. states of Texas, Louisiana, Mississippi, Alabama, and Florida, and other Central American countries.

Halliburton – a major oil servicing company responsible for cementing of the well at the Macondo Prospect.

Hayward, Tony – the beleaguered former Chief Executive Officer of BP that was replaced by Bob Dudley in the aftermath and perceived mismanagement of the Deepwater Horizon oil spill.

Hydrocarbons – molecules that combine carbon and hydrogen that can create large amounts of energy from combustion.

Kickback – a sudden migration of oil, gas, or mud up a well hole.

Liability cap – a limit to the financial liability a defendant will assume. The Oil Pollution Act of 1990 had expanded the cap on economic damages for spills in the U.S. to $75 million if the accident did not exhibit misconduct.

Liner – the outer steel pipe that separates the wellbore hole from the drill shaft or production pipe that can be contained within.

Macondo Prospect – a potentially rich deep water oil deposit in the Mississippi Canyon of the Gulf of Mexico.

Migratory Bird Act – a series of provisions that imposes fines for harm to various migratory birds.

Minerals Management Service – the former agency responsible for leasing federal petroleum and mineral-rich land, and is responsible for subsequent oversight.

Mitsui Offshore Exploration Company – a 10% partner, along with BP and Anadarko, in the exploration and production of the Macondo Prospect.

Mud – a complex liquid that can be mixed to various densities to balance against the pressure of oil and gas in a well.

Mud-Gas Separator (MGS) – a device that can separate gas from mud rising out of a wellbore.

Negative balance – a condition in a well when there is insufficient pressure from above to balance the pressure from oil and gas below.

Negative pressure test – a test that determines if a well has integrity from materials that may migrate from outside the wellbore into the bore shaft.

Obama, U.S. President Barack – the U.S. president that was responsible for oversight of the Deepwater Horizon oil spill.

Occupational Safety and Health Administration – the lead U.S. government agency charged with the protection of workers.

Oil Pollution Act (OPA) – a series of provisions designed to ensure oil pollution is remedied.

Paleozoic Era – a carbon dioxide-rich ere about 300 million years ago that theories suggest produced much of the oil we consume today.

Paraffins – also called alkanes, the primary constituents of petroleum-based fuels.

Peak oil – a date determined when oil consumption exceeds the discovery of known reserves.

Platform – an ocean surface work area that is floating or ocean-floor secured and used to drill for offshore oil.

Positive balance – a condition in which mud from above is of sufficient weight to balance the pressure from oil below a well.

Positive pressure test – a test that pressurizes a well and monitors the well leak down to ensure the well has integrity.

Principal–agent – an economic tool that analyzes whether those charged with representing owners do so in the owners' best interests.

Production casing – a piece of pipe that simultaneously provides a path for oil to the surface and a barrier to the surrounding rock and sand.

Reserves – known pools of oil of a size that is quantified by the estimated amount of oil held within.

Rig – a collection of equipment designed to explore for (exploration rig) or produce oil (production rig).

Riser – the pipe that brings oil from the subsea wellhead to the surface.

Roughneck – a term for workers who drill for oil.

Salazar, Secretary of the Interior Ken – the Obama administration appointed secretary in charge of the Minerals Management Service at the time of the Deepwater Horizon oil spill.

Semisubmersible – a type of oil drilling rig that can be floated to an exploration site and then subsequently partially sunk and anchored to the ocean floor or positioned through complex propulsion systems.

Shell – one of the major oil companies that explore and produce in the Gulf of Mexico.

Shoe – the bottom of an exploratory well designed to seal the well until production.

Special Master – an appointed adjudicator that can assist in streamlining of claims of the court.

Static kill – the sealing of a blown-out well that is balanced and the flow is stopped.

Subeconomic Oil – oil in known reservoirs that cannot be extracted profitably at prevailing oil prices.

Tieback – a type of lower well casing and pipe design that can offer an intermediate seal against oil and gas that might try to migrate through the well in an uncontrolled manner.

Top kill – a cement or metal seal of a well at or near its top.

Transocean – the world's largest oil drilling company commissioned by BP to drill the Macondo Prospect.

Wellbore – the hole drilled in reservoir exploration. A lining is typically inserted into the wellbore to ensure the surrounding substrate does not collapse into the hole.

Wells, Kent – the technical director and BP Vice President that held regular conferences with the press interested in the progress of the well shutdown.

Index